T0186691

C2 Re-envisioned

The Future of the Enterprise

C2 Re-envisioned

The Future of the Enterprise

Marius S. Vassiliou

David S. Alberts

Jonathan R. Agre

CRC Press
Taylor & Francis Group
Boca Raton   London   New York

CRC Press is an imprint of the
Taylor & Francis Group, an **informa** business

CRC Press
Taylor & Francis Group
6000 Broken Sound Parkway NW, Suite 300
Boca Raton, FL 33487-2742

Printed on acid-free paper
Version Date: 20141007

International Standard Book Number-13: 978-1-4665-9580-4 (Hardback)

Visit the Taylor & Francis Web site at
http://www.taylorandfrancis.com

and the CRC Press Web site at
http://www.crcpress.com

To my mother, Avra S. Vassiliou, for everything.

(MSV)

To my wife Bette and my son David, who
provide boundless inspiration.

(DSA)

To Teresa and our three great kids–Elizabeth, Aaron, and Daniel.

(JRA)

Disclaimer

The views expressed are those of the authors and do not reflect the official policy or position of the Institute for Defense Analyses, the U.S. Department of Defense, or the U.S. Government.

Contents

Acknowledgments

We gratefully acknowledge the support and encouragement of David Jakubek and Rob Gold. We also acknowledge Dr. Cynthia Dion-Schwarz, who supported our work in earlier days, and who first suggested writing this book. We acknowledge the support of our employer, the Institute for Defense Analyses.

We are grateful for the research work conducted at the U.S. Department of Defense's Command and Control Research Program throughout the years. The same applies to the North Atlantic Treaty Organization's System and Analysis Study Groups 065 ("C2 Maturity Model") and 085 ("C2 Agility"). These three organizations have contributed greatly to creating a modern body of public knowledge in the area of Command and Control.

This work was supported under the Institute for Defense Analyses Contract No. DASW01-04-C-0003, Task Order AK-2-2701.

1

INTRODUCTION, AND A
TALE OF TWO MISSIONS

1.1 What We Mean by Command and Control

Despite the central importance of Command and Control (C2) in military affairs, there is no universally accepted definition of the phrase. In very general terms, it is usually understood to indicate the management of personnel and resources in the context of a military mission. In this book, we adopt a fairly expansive definition:

> "Command and Control" (C2) denotes the set of organizational and technical attributes and processes by which an enterprise marshals and employs human, physical, and information resources to solve problems and accomplish missions.

This definition can apply to conventional military operations by a single country, coalition military operations by several countries, civil-military partnerships conducting disaster relief operations, civilian law enforcement, and many other endeavors.

Understood this way, Command and Control may seem identical to ordinary management, and in fact the definition above could easily apply to ordinary commercial businesses and their operations. It is thus important to note that the "organizational and technical processes" in a military mission may be very different from those in a commercial business, as are the human, physical, and information resources. Also different are the stakes associated with an operation's outcome. In the military case, lives are at risk, and national security may hang in the balance.

Appendix A discusses other definitions of C2, most of which can be subsumed by the one above.

1.2 What This Book Is About

This book is about the *approaches* that enterprises can take to solving problems and accomplishing missions. It encompasses organizational and technical factors, at a conceptual level. The book discusses and analyzes the large trends that have been altering the C2 landscape in recent years. It also shows how C2 can go wrong and steps that can be taken to reduce the risk of such failures. It argues that the approach to C2—indeed, the approach to enterprise problem solving in general— is an important variable that can be optimized for a given mission and circumstances. Successful enterprises in the future will be those that can reconfigure their approaches in an agile manner to suit the mission and the conditions at hand.

1.3 What This Book Is *Not* About

This book is not about decision theory, course-of-action planning, or communications theory. Although we sometimes touch on such areas, a detailed exposition of them, or other important science and technology topics underlying the practical functions of a C2 system, is not part of our purpose here. This book also makes no attempt to describe, compare, or contrast C2 equipment, systems, and processes used by the U.S. military or other defense establishments. Those seeking detailed information about, say, the Global Command and Control System (GCCS), will not find it here.

1.4 Prelude: A Tale of Two Missions

Before we begin our detailed discussion in the chapters to come, let us take a moment to look at two cases that illustrate some important themes. One case, Admiral Horatio Nelson's victory at Trafalgar, involves a successful mission. The other, the U.S. Hostage Rescue Mission of 1980, involves a failed one.

1.4.1 Nelson at Trafalgar

The Battle of Trafalgar (October 21, 1805) was fought off the southwest coast of Spain by the British Royal Navy, led by Admiral

Horatio Nelson,[*] against a combined fleet of the French and Spanish navies[†] led by French Admiral Pierre-Charles Villeneuve.[‡] The battle was perhaps the most important naval engagement of the War of the Third Coalition of the Napoleonic Wars (1803–1815). The French and Spanish lost 22 of their 33 ships, while the British lost none of their 27.

Nelson achieved victory at least partly by departing from prevailing naval tactical orthodoxy. His departure from such orthodoxy was, in turn, enabled by a change in the approach to Command and Control.

Naval tactics at the time involved engaging an enemy fleet in a single line of battle parallel to the enemy to maximize fields of fire and target areas. Before the development of this widely accepted tactic, fleets had usually engaged in relatively disorganized mêlées, with opposing sides becoming mixed together. Such mêlées made it difficult or impossible for a commander to exercise control. The newer tactics greatly increased the ability to control the engagement because they also facilitated the communications (e.g., via signaling flags) between and among ships necessary to coordinate disengagement from the enemy. Thus, either side could break off at will. This allowed one side or another to limit its losses, and also often led to inconclusive results.

Senior British officers were well aware of the limitations of this way of fighting, and of their communications capabilities (signal flags). Attempts were made to increase the speed of signaling by revising the official signals book, as reflected by the publication of *Popham's Telegraphic Signals; or Marine Vocabulary* in 1803.[§] However, the desire for change did not generally extend to the development of new tactics. Nelson had been thinking about how to overcome existing C2 limitations to achieve a decisive outcome. He understood the inter-relationship between naval tactics and the approach to C2, and recognized the need to co-evolve the two.

Nelson's tactical solution was to divide his fleet in two and attack the enemy line at two points, approaching perpendicular to the enemy

---

[*] 1758 to 1805.

[†] Some of our narrative about Trafalgar is based on an unpublished case study by General Waldo Freeman.

[‡] 1763 to 1806.

[§] Popham (1803).

line, and cutting it into three pieces. He would then surround one-third of the enemy fleet and force a decisive fight. His plan was to cut the enemy line just in front of the flagship, disrupting its communications with many of its sister ships. However, this tactic would also make it impossible for him to stay in touch with all of the ships of his own fleet. Therefore, he also needed a C2 solution that would ensure that the engagement would not result in a lack of control.

Nelson's solution to the C2 problem was to organize his force so that minimal signaling was necessary. To do this, he took advantage of the experience, skill, and initiative of his ship captains. In the months before the battle, Nelson met with his captains to discuss his new tactics. He expected his captains to take initiative within the context of the overall concept, and he impressed upon them the criticality of their role. He wanted to make sure that as the battle unfolded, each captain would know what was expected of him—to understand what to do with his ship to best contribute to the success of the operation without the need for signaling. Nelson followed up on his meetings with his captains by putting the concept into writing. His *Trafalgar Memorandum* was a brief and clear statement of intent that did not attempt to anticipate every eventuality in detail. Nelson recognized the unpredictability of war: "Something must be left to chance; nothing is sure in a Sea Fight beyond all others"[*][†] (see Figure 1.1). Circumstances would dictate execution, subject to adherence to the tactical concept of cutting the enemy line such that a superior force was concentrated on the isolated part of the enemy fleet. In addition to developing shared awareness of intent and the rules of engagement, Nelson fostered tactical awareness by painting his ships in a highly visible yellow and black pattern (later known as the "Nelson Chequer") to make it easier to identify friend from foe. Nelson's trust in the experience, skill, initiative, and shared understanding of his captains enabled him to send his legendary signal, "England expects that every man will do his duty" (see Figure 1.2)

---

[*] Nelson (1805); reproduced in Hodges and Hughes (1936); Kenyon (1910).

[†] Kenyon (1910) reproduction, p. 10. The unpredictability of war, and the need for flexibility of action in the face of it, would also become central in the thinking of the Prussian general von Moltke the Elder, one of the fathers of mission command. We discuss this further in Chapter 6. See also Vincent (2003).

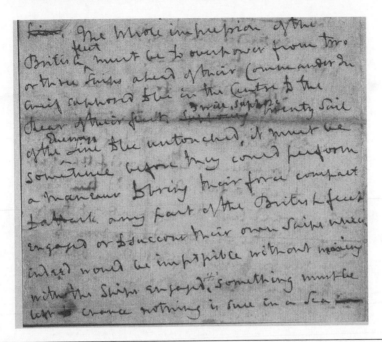

Figure 1.1   A portion of Nelson's original Trafalgar Memorandum of 1805, now held at the British Library. Toward the bottom, we can see the beginning of the sentence "Something must be left to chance; nothing is sure in a Sea Fight beyond all others." (From Nelson, 1805; reproduced in Hodges and Hughes, 1936; Kenyon, 1910. With permission.)

using Popham's flag-based telegraphic vocabulary, and have his fleet win the Battle of Trafalgar, even though he himself was shot and lay dying during part of the engagement.[*]

It is worth noting that trust and shared intent also existed between Nelson and his own superior, Lord Barham.[†] Lord Barham gave Nelson very brief and general instructions in the final phases of the Trafalgar campaign, captured in less than 250 words.[‡] This was only possible because Barham and Nelson had confidence in each other, and knew that they shared the same purpose and strategy. Barham knew he could rely on Nelson to act correctly within their shared framework, using his judgment and experience.

---

[*] Ball (2006); Popham (1803); www.aboutnelson.co.uk (2014); Hodges and Hughes (1936).

[†] Charles Middleton, First Baron Barham, 1726–1813.

[‡] Barham (1805).

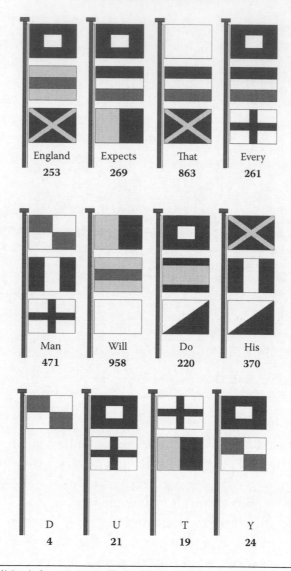

Figure 1.2   Nelson's famous signal, "England Expects That Every Man Will Do His Duty," using Popham's telegraphic vocabulary.

Nelson's victory at Trafalgar shows that a change in the then *standard* approach to C2 was required to support a desired change in warfighting tactics. The changes made to the C2 Approach involved a different delegation of decision rights, one with more freedom of action given to subordinates. Instead of an accompanying increase in communication bandwidth (faster signaling), Nelson realized that developing *a priori* shared awareness, coupled with improved friend-and-foe identification capability, would enable self-synchronization. Thus, his captains could operate without the usual amount of information exchange.[*] To make this new C2 Approach work in practice required developing in advance a shared understanding of both the new tactics and the new C2 Approach. This illustrates the importance of co-evolution in doctrine, organization (C2 Approach), and materiel that are a central tenet of networked enabled operations. The main point is that mission effectiveness was dramatically increased when a new tactic, enabled by a change in the C2 Approach, was adopted.

1.4.2 U.S. Hostage Rescue Mission of 1980

During the period of April 24 to 25, 1980, the United States mounted a mission to rescue hostages being held in the U.S. embassy in Tehran, Iran. The mission failed, with eight deaths among the involved personnel.

During the mission, a C-130 transport airplane heading to the rendezvous landing site ("Desert One") encountered a large desert dust cloud (known in Iran as a *haboob*). The *haboob* was not a major problem for the airplane, but it was potentially a serious threat to the eight helicopters following far behind it. The airplane did not warn the helicopters because of a strict dictate of radio silence. There was a chance the aircrew could have used a secure satellite radio to issue the warning, but unfamiliarity with the equipment made them unable to work out the coding parameters.[†]

The helicopters thus entered the *haboob*. Because of radio silence, they could not tell each other what they were doing or where they were going. One helicopter had to abort because of a suspected blade failure, and two others left the *haboob* and landed. One of the two that landed prematurely was that of the group's leader. The leader made a secure call to a U.S. command center in Egypt and was told to proceed to the rendezvous landing site ("Desert One"), but none of the other helicopters could hear the conversation. The other pilot who had landed prematurely was no longer in visual contact.

---

[*] This finding was echoed in the early days of the conceptual development of net-centric warfare. Contrary to some expectations bandwidth utilization can sometimes decrease when operations are "network enabled."

[†] Anno and Einspahr (1988); Bowden (2006).

Because of readings indicating malfunctions and the difficulty of flying again through the *haboob*, he made an independent decision to return to the aircraft carrier *Nimitz*. To make things worse, his was the helicopter carrying all the spare parts needed for possible repairs. None of the helicopters could talk directly to Desert One and thereby learn that the rendezvous landing site was clear. Later, the pilot who returned said he would have continued had he known that fact.[*] The inability to communicate led to the loss of needed helicopters and crucial spare parts at Desert One. The mission was canceled on the ground after several other missteps, and during the retreat one of the helicopters collided with one of the transport planes, killing eight soldiers (see Figure 1.3).

The failed mission had a number of organizational and structural problems. It involved U.S. Army Delta Force, U.S. Army Rangers, U.S. Air Force pilots, and U.S. Navy helicopter pilots, among others, in a highly complex operation. The mission was adversely impacted by an inadequate approach to C2 that suffered from compartmentalization and evidenced mutual distrust between and among these service components. There was also a lack of unified command, with no single component commander to unify the Air Force airplanes and Navy helicopters, and no single ground commander to unify Delta Force and the Rangers.[†]

In addition, there were other communication problems besides the ones originating from security constraints. The Army Rangers who were guarding the landing site in the Iranian desert used radios that could not communicate with Delta Force or Air Force personnel. They were also unable to inform ground commanders in a timely fashion when a bus full of Iranian civilians appeared, complicating the operation. The landing site could not talk to the helicopter fleet.[‡] In other words, there was a pervasive and crucial lack of interoperability. Expensive communications equipment is of little use if it cannot talk to other communications equipment.

---

[*] Anno and Einspahr (1988).
[†] Ibid.; Gass (1992); Holloway (1980); Thomas (1987).
[‡] Anno and Einspahr (1988).

(a)

(b)

Figure 1.3   The Iran hostage rescue mission of 1980. (a) Sikorsky RH-53D helicopters taking off from the aircraft carrier *Nimitz*, April 24, 1980. (From the U.S. Department of Defense, U.S. Defense Information System photograph ID DN-SC-82-00720; http://commons.wikimedia.org/wiki/File:RH-53Ds_taking_off_from_USS_Nimitz_1980.JPEG. With permission.) (b) Wreckage at the Desert One landing site in Iran. (From the U.S. Department of Defense, http://commons.wikimedia.org/wiki/File:Eagle_Claw_wrecks_at_Desert_One_April_1980.jpg. With permission.)

1.4.3 What Can We Learn?

The two cases above illustrate the stark difference between an operation with shared intent, an adequate level of shared awareness, and an appropriate approach to C2—and an operation without any of those things. In the chapters that follow, we will explore the interplay between organization design, C2 Approach, communications technology, and success or failure. Chapter 7 in particular analyzes the things that can go wrong with C2, adding many more cases to that of the hostage rescue mission discussed above.

2

MEGATRENDS AFFECTING THE FUTURE OF THE ENTERPRISE

In 1982, futurist John Naisbitt published a book entitled *Megatrends: Ten New Directions Transforming our Lives.*[*] In it he identified several long-term, large-scale socioeconomic, environmental, and technological forces (megatrends) that he argued would profoundly affect the future. He was substantially correct in his arguments and predictions. Many of his megatrends continue in some form into the present day and are still important influences in our world. Among the megatrends that Naisbitt discussed in 1982, were the change from an industrial society to an information society, economic globalization, and various organizational tendencies.

The concept of a megatrend has since entered common discourse, with many authors and organizations identifying various large-scale forces they think are shaping the future.[†] We believe that today, the state of the art and practice of C2, the nature of the missions we undertake, and the environment in which these missions are undertaken are being continuously molded by four megatrends that we discuss below. Some are related to Naisbitt's original megatrends from 30 years ago.

2.1 The Four Megatrends

The four megatrends, depicted in Figure 2.1, are as follows:

- *Megatrend 1: Big Problems*—This is manifested in part as increasing complexity of both endeavors and enterprises, as military establishments form coalitions with each other and partnerships with various civilian agencies and nongovernmental organizations.

---

[*] Naisbitt (1982).
[†] For example, Hajkowicz et al. (2012); Singh (2011); PWC (2013); Vassiliou and Alberts (2012).

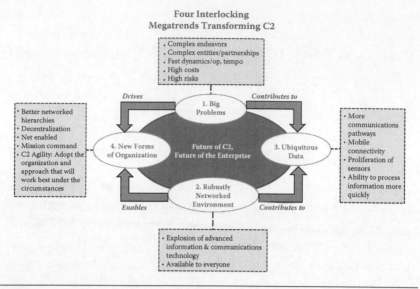

Figure 2.1   The four megatrends determining the future of C2 and the future of the enterprise. (Modified from Vassiliou and Alberts, 2012.)

- *Megatrend 2: Robustly Networked Environments*—These are enabled by the extremely broad availability of advanced information and communications technologies (ICTs) that place unprecedented powers of information creation, processing, and distribution in the hands of almost anyone who wants them—friend and foe alike. This is related to one of Naisbitt's original megatrends, the transition from an industrial society to an information society.
- *Megatrend 3: Ubiquitous Data*—This refers to the unprecedented volumes of raw and processed information with which human actors and C2 systems must contend.
- *Megatrend 4: New Forms of Organization*—Decentralized, net-enabled approaches to C2 have been made more feasible at least in part by Megatrend 2. This is related to other Naisbitt megatrends, such as a transition from hierarchies to networks and a movement from centralization to decentralization in both politics and business.

These megatrends interact with each other in various ways. Big problems and complex endeavors often demand complex enterprises, which often operate more effectively in a decentralized, net-enabled manner.

Decentralized complex enterprises often generate larger volumes of information than hierarchical ones, contributing to Ubiquitous Data. Meanwhile, the Robustly Networked Environment and advanced information and communications technology enable and facilitate both complex endeavors and net-enabled decentralized approaches. They also empower more actors to create more information, and thus contribute directly to Ubiquitous Data.

The four megatrends demand changes in the ways we interact and work with one another while, at the same time, providing us with opportunities to make the necessary changes to our institutions. Such changes can improve the likelihood of success in the face of increasing complexity and dynamics. Thus, these trends are determining winners and losers. The winners are those institutions that adapt. Enterprise adaptation, for the purposes of this book, involves how enterprises allocate decision rights, facilitate and/or restrict interactions, and disseminate information. This is discussed in more detail in subsequent chapters. As the Information Age has matured and we, as individuals and institutions, are increasingly connected at all times and wherever we happen to be, an "Age of Interactions" has arisen. In the new age, we are witnessing changes that profoundly affect what information is available and how we seek and receive information. This is affecting every human endeavor, both public and private, from warfare, to commerce, to education, to interpersonal relations. Powerful and affordable new technologies are enabling us to rethink how we organize, providing us with virtually limitless access to information and to one another, and are greatly increasing the tempo of events. As a result, the pace at which we work and live is often increasing faster than our ability to cope.

Today's organizations often struggle because they have been slow to recognize that these developments have created a need for structural change—that is, changes to the nature of entities and how they function. Many institutions have responded by changing only at the margins in a futile effort to preserve existing equities and the ways decision rights are currently allocated. Simply put, those individuals and organizations that are resisting change are becoming maladapted for the world in which we live and must operate.

In the next four chapters we will discuss and analyze each of the four megatrends in turn.

3
MEGATREND 1: BIG PROBLEMS

Why are traditional approaches to problem solving and organization increasingly leaving us wanting as we try to grapple with many of the problems we face? One of the reasons is that the problems we need to solve are actually becoming more challenging. Some have called these increasingly difficult problems "wicked problems."[*] Others think in terms of what it takes to solve such problems, and refer to efforts to tackle them as complex endeavors.[†]

What makes some situations or tasks so daunting, and some problems so difficult to solve? Answering this question can help us understand why our existing institutions, familiar approaches, and available tools might be inadequate. It can also point us to ways of avoiding or coping with these situations, accomplishing these challenging tasks, and solving these seemingly intractable problems. Success requires both knowing what to do and being able to do it. As we see, some of our difficulties lie in not being able to formulate a solution strategy. In other cases, we know what to do, but we cannot actually do it. There are also times when we know what to do and it is possible to do it, but we cannot organize effectively and efficiently enough to get the job done.

Solutions are not really solutions unless they are both feasible and affordable. Feasibility and affordability depend on the nature of the particular entity in question. Difficulty is a function of both the characteristics of the problem and the capabilities of an entity. Problems become less and less tractable when entities are less than well adapted and resourced. Big Problems stress the ability of entities to obtain and bring to bear the resources they require in a timely

---

[*] Consider, for example, Rittel and Webber (1973).
[†] Alberts and Hayes (2007).

manner. Resources include not only materiel (e.g., facilities, equipment, systems, and supplies) but also information and time. A failure to resolve Big Problems presents significant risks, including the risk of catastrophic failure.

3.1 Difficulty of Big Problems

Big Problems possess a set of characteristics that make them difficult for traditional organizations to resolve using traditional methods. These characteristics include size, dynamics, complexity (mission and enterprise), and the cost of error. As shown in Figure 3.1, size is related to affordability, dynamics to time pressures, complexity to uncertainty, and the cost of error to risk. The four megatrends are interrelated. Robustly Networked Environments, Ubiquitous Data, and New Forms of Organization affect, and are affected by, increasing problem difficulty. On the one hand, each of these three contributes to the factors that make problems more challenging, while on the other hand, each of the three provides opportunities for entities to adapt in ways that make them better equipped to handle the challenges associated with Big Problems.

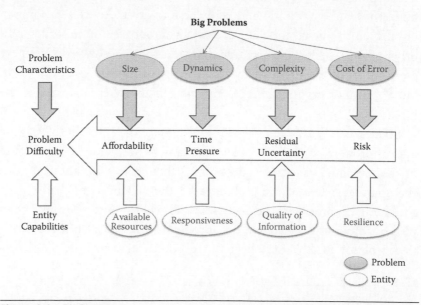

Figure 3.1 Big Problems.

3.2 Size and Affordability

The affordability of a solution depends on the resources of an entity compared to those demanded by the size of the problem. A large problem size, and the resulting lack of affordability, can drive organizational complexity. Since the very size of a Big Problem may tax the available resources of even the largest and most powerful entity, it may require assembling a collection of heterogeneous organizations to make the requisite resources and skills available. However, even when the combined resources of a collection of entities are available, success depends on the ability of this collective enterprise to come together as an effective entity in a timely manner so that all participating organizations can synchronize their individual efforts.

Companies outsource or create and participate in supply chains and ecosystems while militaries seek allies or form coalitions. These coalitions may also expand to include collaboration with civilian entities, including law enforcement and nongovernmental organizations. For example, Hurricane Katrina, which struck the American state of Louisiana on August 29, 2005, required a multidimensional response involving hundreds of military and civilian entities.[*] The act of moving from a single entity to a collection of entities (i.e., the creation of an enterprise) presents immediate challenges in management, governance, and command and control. Who is in charge? How are priorities and allocations of resources determined? How are conflicts of interest resolved? What capabilities and processes exist to share information and work across entity boundaries? How are actions coordinated and synchronized?

What is needed is an approach to the enterprise that transforms the parts into an effective and efficient whole. Less well recognized but also critical is the adoption of an approach by each participating entity that is designed with the collective enterprise in mind.

3.3 Dynamics and Time Pressure

The dynamics of a situation affect the time pressure imposed on a response. The ability to respond in a timely manner depends on how

---

[*] Moynihan (2009); see also Chapter 7, this volume.

long it takes to recognize the situation, understand it, take appropriate action, and have the anticipated effects of the action manifested and the situation altered in a desirable way. Even while an entity or collection of entities is in the process of responding, situations can change. The rate of change can also change and as a result reduce or increase time pressure. The rate of change in the situation can, if it is rapid enough, make the actions that entities take less effective or even counterproductive. Thus, even entities that are capable of responding quickly may find that they are unable to keep up. As a result, the situation can spin out of control, resulting in failure.

The time available for entities to respond has been reduced in recent years. This shrinking window of opportunity is a result of increasing problem complexity, enterprise complexity, and the Robustly Networked Environment (Megatrend 2). The increased connectedness associated with the Robustly Networked Environment has been enabled by advances in information and communications technology (ICT). These same advances have, in turn, enabled the creation of new Enterprise Approaches. Figure 3.2 identifies some of the reasons why time pressure has been increasing. As the world has become more connected, interactions that never would have occurred previously are now possible, and even probable. Cascades of consequences can occur

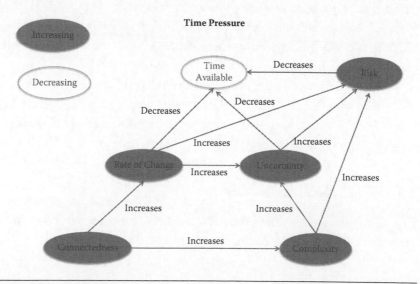

Figure 3.2   Time pressure.

with greater frequency. These, in turn, translate into increased complexity that adversely impacts one's ability to make predictions. Thus, while size and affordability contribute to *organizational complexity*, time pressure contributes to *problem complexity*.

Increased dimensionality is also a major contributor to problem complexity. It has become routine to face situations that require understanding of multiple disciplines and the interactions between and among them. This serves to make our ability to understand these situations more difficult and make solutions elusive, as second- and third-order cross-dimensional effects can make analysis time consuming and nullify what may seem like progress in one dimension. Increased complexity is also associated with an increase in the likelihood of non-linear effects—small changes in the value of one variable having disproportional impacts on others. Taken together, these factors make situations and actions far less predictable and increase the probability of incurring significant losses or costs. Both the increased risk and lack of predictability make it more important to resolve situations more expeditiously lest they quickly deteriorate and become even more difficult and costly to resolve. This urgency increases time pressure.

With the advent of Ubiquitous Data (Megatrend 3), more and more information is being sensed and disseminated more rapidly (due to increased availability of communications pathways, mobile connectivity, the proliferation of sensors, and the ability to process information more quickly), and entities have the information they need to act more quickly, but only if they have the ability to handle the increased workload. New Enterprise Approaches, also enabled by the Robustly Networked Environment and advanced ICTs, feature streamlined, agent-assisted, and automated processes that allow entities to decide and act quickly. This increases the tempo of action and reaction, leading to faster feedback loops and creating even more time pressure.

3.4 Problem Complexity and Residual Uncertainty

Problem complexity can have the effect of increasing uncertainty. Uncertainty, in turn, affects our ability to resolve problems in a number of ways:

- A lack of certainty adversely affects our ability to choose an appropriate response to a problem or have confidence in our selection.

- A lack of certainty can lead to putting off decisions until more information is collected and processed, thus delaying decisions and actions.
- A lack of confidence can result in selecting a tentative approach.
- Uncertainty can result in micromanagement (selection of more centralized approach options).
- Uncertainty increases cognitive load.

All four megatrends contribute to increasing uncertainty:

- Robustly Networked Environments are enabling an explosion in the number of connections, increasing instability and unpredictability.
- Ubiquitous Data, while offering opportunities to increase understanding and reduce uncertainty, can actually result in creating more uncertainty in some individuals and organizations. The ability to collect so many facts may lead to a perpetual feeling that not all the relevant ones have been collected, and that with just a few more, we can make a "perfect" decision.
- New Enterprise Approaches that, because of the nature of the problem, require a large collective that is composed of heterogeneous entities, create uncertainty because members are not familiar with all aspects of the problem, are not familiar with each other, have different ways of organizing and operating, and may not communicate and exchange enough information with one another.

3.5 Risk

Risk is a function of both the probability of an event and its expected cost. In deciding how to respond to a situation or to work toward solving a problem, decision makers identify and assess a set of options. An important part of this process involves predicting the consequences of different options under different circumstances. An estimate of risk is an important part of this process.

There are a variety of sources of risk that need to be considered. First, there is, for each option under each of a set of potential circumstances, the risk of cascades of consequences in multiple domains that would have adverse impacts on the situation or the entity. This can be thought of as collateral damage. Second, there are risks associated with an entity's ability to actually execute the option.

To the extent that entities depend upon ICT to execute an option, any degradation or disruption to ICT-related services will increase risk. This risk needs to be balanced against the risks associated with being forced to choose among options that do not leverage ICT or to adopt Enterprise Approaches that are not enabled by ICT. In other words, as ICT advances it offers us opportunities that reduce operational risk while potentially increasing exposure to technical risk.

3.6 Big Problem Stresses on Traditional Enterprises

Big Problems, by virtue of their size, dynamics, and the uncertainty and risks associated with them, are making it increasingly difficult for traditional organizations to maintain their effectiveness and efficiency—the reasons that they have become successful in their competitive spaces.

Clearly, undertakings of enormous scope, requiring large amounts of resources, have been successfully undertaken. There are many examples, including construction projects such as the Panama Canal, transcontinental railways, and the Chunnel; exploration of the poles, the oceans, the moon, and more recently outer space; and scientific advances such as the sequencing of the human genome, and the characterizing of subatomic particles. Large commercial ventures have likewise proven to be within the realm of the possible. However, our success in these endeavors has not translated into success in confronting the challenges of terrorism, failed states, climate change, hunger, economic instability, disaster relief, and governance.

Military, industry, and government organizations, as well as individuals, are currently struggling with the reality of cyberattacks of increasing frequency and severity. Attacks, and threats thereof, have adversely affected our ability to fully utilize the power of information and networking capabilities. As a consequence, Big Problems become even more difficult to resolve.

Traditional organizations are well adapted for challenges for which they have adequate resources and that have the following characteristics:

- Decomposability—that is, various parts of the problem can be handled separately and in relative isolation

- Familiarity
- Stability
- Predictability

These characteristics permit organizations to specialize functionally, develop competence, incrementally improve operations, and plan for such challenges. However, when resources are inadequate or one or more of these characteristics are no longer present, traditional organizations may need to adapt how they are organized and how they operate.

The availability and feasibility of alternative forms of organization (Megatrend 4) are discussed in Chapter 6, which considers ways to characterize such alternative organizations. In Chapter 8, we provide evidence that more "net-enabled" and decentralized organizations, with broader allocation of decision rights, broader interaction patterns, and broader distribution of information, can be more effective for many problems and circumstances. Chapter 8 will also provide evidence and arguments that *agility*, the ability to choose the appropriate approach depending on the mission and the circumstance, is crucial to enterprise performance.

First, however, we consider the other two megatrends.

4

MEGATREND 2: ROBUSTLY NETWORKED ENVIRONMENTS

4.1 Explosion in Advanced Information and Communications Technologies

The rapid advancement of information and communications technologies (ICTs) has placed unprecedented computing power, information storage, information processing, and connectivity into the hands of almost anyone who wants it. Such technologies include personal computers and peripherals, networking, geolocation, and mobile telephony.

When one lives in an era of rapid progress, the changes can appear almost commonplace, and it is easy to take them for granted and forget their profundity. It is thus helpful to review some basic statistics. One useful case to consider is mobile telephony. Perhaps no other technology represents so many facets of the revolution in ICT: computing power, multimedia, storage, and connectivity, with both hardware and software aspects being crucial. In 2001, there were 15.5 mobile cellular subscriptions for every 100 people in the world.[*] In 2013, there were 96.2.[†] The growth was particularly striking in the developing world,[‡,§] which went from only 7.9 cellular subscriptions per 100 people in 2001, to 89 in 2013.[¶] Meanwhile, the developed world had, by 2013, more cellular subscriptions than people (127 per

---

[*] ITU (2011).

[†] ITU (2013a).

[‡] Definitions of *developed* and *developing* world are those used by the United Nations (2012).

[§] OECD (2011).

[¶] ITU (2011, 2013a).

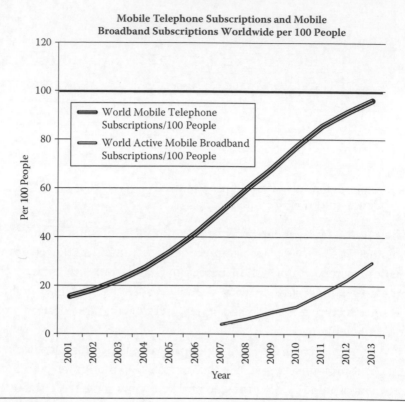

Figure 4.1   World mobile telephone subscriptions and active mobile broadband subscriptions per 100 people.

100 people, compared to 47.1 in 2001).* Figures 4.1 and 4.2 show the trends in mobile telephone subscriptions in various countries. In 2003, 61% of the world's population resided in an area with cellular coverage. By 2010 the figure was 90%.† This all represents a truly staggering growth in communications power and connectivity for individuals and organizations everywhere.

Not only has mobile telephony exploded in numbers of handsets and subscriptions, but the character of the handsets and the range of available services have also radically improved. Smartphones, that is, handsets capable of accessing the Internet, producing and receiving imagery and video, and a myriad other functions, have gone from being elite items a few short years ago to becoming the dominant

---

* ITU (2013a).
† ITU (2012).

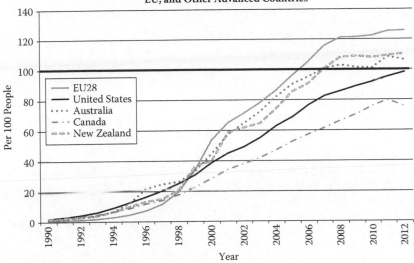

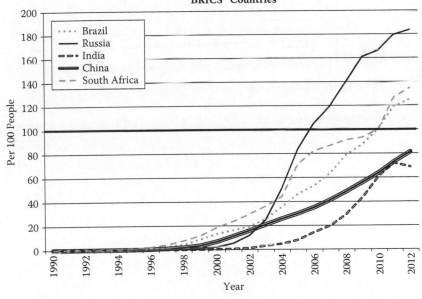

Figure 4.2 Mobile telephone subscriptions per 100 people in selected western countries, the BRICS countries (Brazil, Russia, India, China, and South Africa), and some other countries of interest. (*continued*)

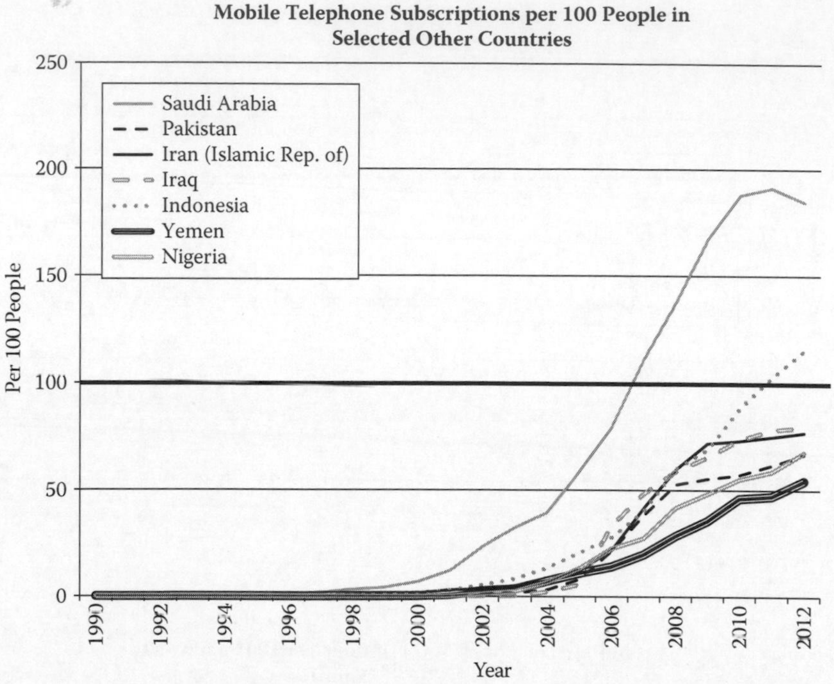

**Mobile Telephone Subscriptions per 100 People in
Selected Other Countries**

Figure 4.2   (*continued*) Mobile telephone subscriptions per 100 people in selected western countries, the BRICS countries (Brazil, Russia, India, China, and South Africa), and some other countries of interest.

handset in many areas. In the United States, the proportion of cellular phones that were smartphones was less than 10% in 2008.[*] By the third quarter of 2011, 65% of all new handsets being shipped in the United States were smartphones.[†] Globally, the sales of smartphones exceeded those of conventional cellular phones in 2013.[‡] The United States had 74.7 mobile broadband subscriptions per 100 people in 2012, a nearly ninefold increase over just four years earlier.[§] Globally, the figures were lower, but still significant: mobile broadband subscriptions stood at 29.5 per 100 people in 2013.[¶] In the developed world, the number was 74.8 and in the developing world 19.8.[**] Figure 4.3

---

[*] Kwoh (2011).
[†] Vision Mobile (2011).
[‡] Gartner (2014).
[§] ITU (2013b).
[¶] ITU (2013a).
[**] Ibid.

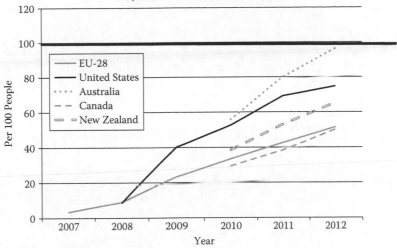

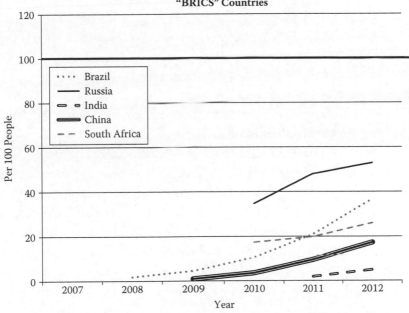

Figure 4.3 Active mobile broadband subscriptions per 100 people in selected western countries, the BRICS countries (Brazil, Russia, India, China, and South Africa), and some other countries of interest. (*continued*)

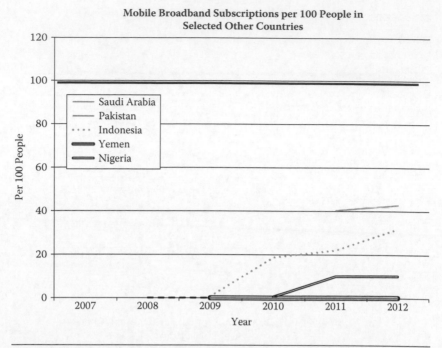

Figure 4.3   (*continued*) Active mobile broadband subscriptions per 100 people in selected western countries, the BRICS countries (Brazil, Russia, India, China, and South Africa), and some other countries of interest.

shows the trends in active mobile broadband subscriptions in various countries. Data resulting just from mobile phones grew 81% in 2013, reaching 1.5 exabytes per month.[*][†]

To put the power of smartphones in perspective, compare them to computers of the past. The LINPACK[‡] benchmarks for numerical computation test the speed of a computer for a particular set of matrix calculations.[§] While LINPACK results are by no means a measure of total system performance, they are a useful indicator of one important dimension of computing power, and they also have the advantage that they have been recorded for decades. A Cray 1 supercomputer—the

---

[*] Cisco (2014).

[†] (See Appendix B, this volume.) An exabyte is 10^{18} bytes, or a billion gigabytes, if using the decimal scale. If using the binary scale, an exabyte is 2^{60} bytes, or about 1.153×10^{18} bytes. Unfortunately, most sources do not specify which definition they are following. General-interest literature usually uses the decimal scale.

[‡] LINear equations software PACKage.

[§] Dongarra et al. (1979); Dongarra et al. (2001); Dongarra (2007, 2011).

fastest computer in the world in 1979—had a performance of about 3.4 MFLOPS.[*,†] A Cray XM-P-4 supercomputer from 1986 had a performance of about 220 MFLOPS per processor.[‡] Some of the most recently benchmarked smartphones exceeded the latter number. The fastest phone benchmarked in 2012 had a performance of 258 MFLOPS.[§,¶] In some newer tests, one smartphone with a multicore processor turned in a performance of 316 to 709 MFLOPS, depending on test conditions.[**]

As for bandwidth, the current "fourth-generation" or "4G" cellular technology, as exemplified by the LTE (Long-Term Evolution) specification, is capable of average data rates of 8 to 20 Mbps[††] or more, depending on the particular implementation.[‡‡] The previous generation of service, "3G" (as exemplified by the ITU's[§§] IMT-2000 specification[¶¶]) was capable of average data rates of 1-2 Mbps before the various enhancements that eventually led to 4G.[***] The ITU's vision for 4G is 100 Mbps in high-mobility applications, and 1 Gbps[†††] in low-mobility ones.[‡‡‡]

Other indicators of the explosion in commercial ICT are not difficult to find. Costs of hard-drive data storage have dropped from around $700 per gigabyte in 1995 to as low as 7 cents per gigabyte in 2009.[§§§] At the end of 1996, less than 1% of the world's people used

---

[*] MFLOPS = 1 million, or sometimes 2^{20}, floating-point operations per second.

[†] Dongarra (2007).

[‡] Ibid.

[§] Greene Computing (2012).

[¶] To the best of our knowledge, these benchmark figures for various devices are roughly comparable. However, one must still exercise caution, since there are a number of different LINPACK benchmarks, corresponding to different matrix sizes, and not all investigators always quote which one they are applying. Also, some investigators report observed performance while others calculate theoretical peak performance over one computational cycle, and not all investigators specify which number they are quoting.

[**] Bennett (2013).

[††] Mbps = million bits per second if using the decimal scale.

[‡‡] Plumb (2012).

[§§] ITU, International Telecommunications Union.

[¶¶] IMT, International Mobile Telephony; see ITU (2003).

[***] ITU (2003).

[†††] Gbps = billion bits per second if using the decimal scale.

[‡‡‡] ITU (2008).

[§§§] Komorowski (2011).

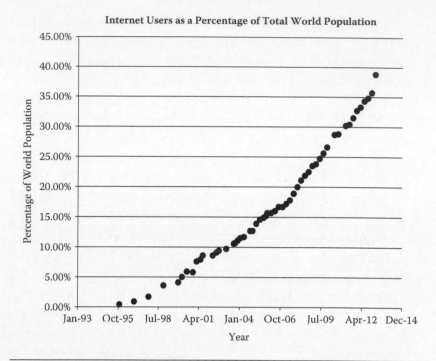

Internet Users as a Percentage of Total World Population

Figure 4.4   Trends in global Internet use. (*continued*)

the Internet (Figure 4.4).[*] By 2013 the number was 38.8% (76.8% in the developed world, and 30.7% in the developing world).[†] The global Internet went from moving an aggregate of about 1 exabyte of data per month in 2004 to 21 exabytes of data per month in 2010.[‡] Global annual Internet Protocol (IP) traffic will likely reach a zettabyte[§] in 2015 and the annual global IP traffic will surpass the zettabyte threshold (1.4 zettabytes) by the end of 2017.[¶] The social networking site *Facebook*, virtually unknown outside Harvard University at the beginning of 2004, had about 1.2 billion regular (at least monthly) users by the end of 2013—around 17% of the world's population (Figure 4.5).[**] According to Intel Corporation, an average *minute* on

---

[*] Internet World Stats (2014).

[†] ITU (2013a).

[‡] Miller (2010); Cisco (2012).

[§] 2^{50} Bytes in the binary system, 10^{21} bytes in the decimal system. (See Appendix B, this volume.)

[¶] Cisco (2013).

[**] Internet World Stats (2012).

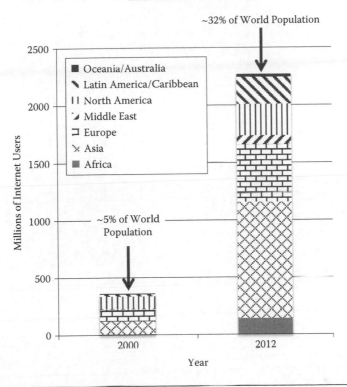

Figure 4.4 *(continued)* Trends in global Internet use.

the Internet sees 640 terabytes of data transmitted, with 204 million e-mails, 6 million Facebook views, 30 hours of video uploads, 1.3 million video views, and 2 million Google searches.[*] Intel also projects that by 2015 the number of networked devices will be twice the world population.

Businesses, governments, and other organizations are heavy users of networking technology, both internally and externally. The Organization for Economic Cooperation and Development (OECD) estimates that by 2011, 96% of all companies in OECD countries were connected to the Internet.[†] Large numbers of organizations also have intranets. The OECD reports that nearly 35% of companies in the European Union and over 20% of Canadian companies had intranets

---

[*] Intel Corporation (2014).
[†] OECD (2012b).

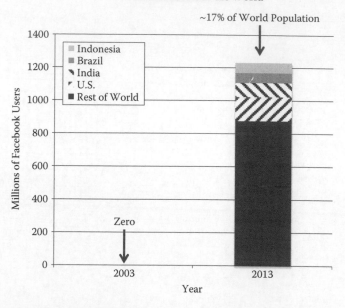

Figure 4.5   Facebook users in the world.

in 2010.* Up-to-date statistics specifically for the United States are difficult to find. A recent survey reports that 85% of the organizations in the International Association for Human Resource Information Management claim to have some sort of intranet or employee portal.† Many or most of these may not be intranets in the sense of the OECD statistics. The U.S. Department of Defense has a large internal network, the Defense Information Systems Network (DISN), whose total capacity grew by a factor of 12 between 2005 and 2010.‡

Organization intranets and internal social networks have sometimes been criticized for not fulfilling their potential in truly tying their entities together.§ Intranets have sometimes been used as one-way information channels to spread company policy to employees in a hierarchical fashion.¶ More recently, they have been evolving into tools to create more closely networked organizations. A cul-

---

* OECD (2012b).
† Appirio (2013).
‡ Moran (2011).
§ Ruppel and Harrington (2001); Martini et al. (2009).
¶ Martini et al. (2009).

ture of innovation and mutual trust increases the chances of success.[*]
Essentially, businesses and other entities now have the technology, if
not always the culture, to adopt a number of innovative and flexible
organizational forms that might hitherto have been unworkable.

4.2 The Technology Explosion Has Been Largely Commercial

While many of today's advanced ICTs have ultimate military origins,
the discussion in the previous section underscores that the tremendous
progress in ICT—particularly in mobile communications, but also in
many other areas—in the last two decades has been in the commercial
arena. It was not always so: before 1990, military-developed technologies often drove commercial applications. However, as early as
1999, a study by the RAND Corporation noted that the U.S. military market accounted for only 2 percent of the demand for information technology in the United States, whereas in 1975 it had been 25
percent.[†] The driving power of commercial technology development
can also be illustrated by some business statistics. In the countries
of the OECD, which include the United States and other advanced
economies, revenues from mobile networks were about $527 billion
in 2009, up from $182 billion a decade earlier.[‡] Industrial research
and development spending on ICTs was about $279.1 billion world-
wide and $148.5 billion in the United States. The research budgets
of five major companies (Microsoft, Cisco, Google, Qualcomm, and
Apple) totaled $26.5 billion in 2011.[§] By contrast, U.S. Department of
Defense spending on research and advanced technology development
in areas contributing to the Net-Centric Joint Capability Area (JCA),[¶]
encompassing communications and networking, was estimated
at $0.385 billion for 2012.[**] The entire U.S. market for Command,
Control, Communications, Computers, Intelligence, Surveillance,
and Reconnaissance (C4ISR) equipment and technologies, including
electronic warfare, was about $43 billion in 2012.[††]

---

[*] Ruppel and Harrington (2001); Martini et al. (2009).
[†] Khalilzad et al. (1999).
[‡] OECD (2012a).
[§] R&D (2012).
[¶] See Appendix A, this volume.
[**] Kramer et al. (2012).
[††] McHale (2012).

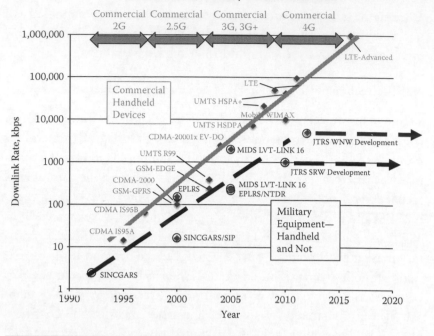

Figure 4.6 Commercial and military wireless data rates.

4.3 Are Commercial ICTs Better?

Consider Figure 4.6, which compares data rates available in commercial and military wireless communications technologies. The U.S. military still uses some SINCGARS radios, first introduced in 1988.[*] Even after improvements, these radios are only capable of data rates up to 16 kbps.[†] The Soldier Radio Waveform of the Joint Tactical Radio System (JTRS) is supposed to be able to achieve 1 Mbps.[‡] JTRS's Wideband Networking Waveform (WNW) may achieve up to 5 Mbps.[§] Commercial technologies, on the other hand, can already provide an order of magnitude more than the higher number and are on track to achieve advantages of two or more orders of magnitude. Does this mean that commercial technology is superior? Perhaps it is, in some dimensions, under some conditions, for some purposes. We should bear

---

[*] Kagan (1999).

[†] Joint Integrated Test Center (JITC) (2012).

[‡] Ibid.

[§] Ibid.

in mind that some military requirements are more stringent. Examples of such requirements are security and resistance to jamming.

4.4 Implications for Military C2

4.4.1 Implication 1: Advanced ICT Is Available to Everyone

The computing and connectivity technologies essential to C2 systems are available to almost anyone, friends and foes alike. Nonstate actors and governments can take advantage of them. A recent study has discovered that foreign fighters in the Syrian Civil War that began in 2011 have made extensive use of social media such as Twitter on mobile devices, both to acquire current information about the conflict and to document their own activities.[*] Under some circumstances, such adversaries may be able to use the afforded capabilities to outmaneuver a large, modern military that uses legacy ICT. We discuss some examples of this in Chapter 6.

4.4.2 Implication 2: Advanced ICT and the Robustly Networked Environment Enable New Forms of Organization

The drastically increased connectivity can allow net-enabled organizations to operate in a more decentralized and flexible manner, while still maintaining shared awareness and understanding. Decentralized, net-enabled organizations are discussed in Chapters 6 and 8.

4.4.3 Implication 3: New Recruits Are Already Immersed in the New Technologies

The young people who are joining today's armed forces are likely to be completely immersed in advanced commercial ICT; to this generation of "digital natives," legacy C2 systems and their associated technologies and policies may seem ineffective, and perhaps almost quaint. For these young people, creating, processing, transmitting, and sharing large amounts of multimedia information are matters of routine.

---

[*] Carter et al. (2014).

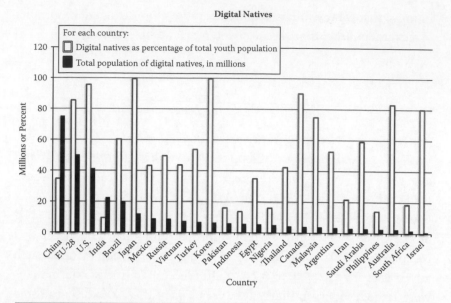

Figure 4.7   Digital natives.

The ITU did a study and analysis of "digital natives" around the world.[*] According to Prensky, who coined the term, a digital native is someone who has grown up immersed in the technologies associated with computers, video games, and the Internet.[†] E-mail, cellular phones, smartphones, and social networking are integral parts of a digital native's life. For its analysis, the ITU adopted a more specific definition of a digital native as a youth between the ages of 15 and 24 with five or more years of experience using the Internet. Using this definition, the ITU estimated that in 2012 there were about 363 million digital natives throughout the world, representing about 30% of the world's youth and 5.2% of the world's total population. In the developed world, digital natives were about 10% of the total population and 89.1% of the youth population, while in the developing world they were 4.2% of the total population and 22.8% of the youth population.

Figure 4.7, calculated and drawn from ITU data, shows the countries with the largest populations of digital natives. Only in the United States and the EU-28 is there a confluence of a large total population and a large proportion of digital natives among the youth populations.

---

[*] ITU (2013a).
[†] Prensky (2001a,b).

This puts these two entities in the front rank of total digital native population. China still exceeds them both, by virtue of its huge total population, despite a much lower proportion of digital natives among its youth (34.7% for China, versus 85.4% for the EU-28 and 95.6% for the United States). Other countries like Korea and Japan have even higher proportions of digital natives among their youth (99.6% and 99.5%, respectively), but their lower total populations yield lower absolute numbers.

4.4.4 Implication 4: Military Establishments Must Adopt Commercial ICTs When Possible

Can military establishments adopt commercial ICTs? In some cases, the answer is "yes," although there are some limitations. Certainly military establishments have adopted modern cellular and networking technologies in their business operations, and for more routine situations. At the tactical edge, however, problems arise from the special requirements of the warfighters and involve environmental, technical, policy, and acquisition considerations. Typical environmental examples are survivability in the face of hostile action, lack of fixed infrastructure, high mobility, and ruggedness. Technical issues include robustness (in the face of loss of signals) and security. Policy and acquisition impediments arise from the myriad regulations and processes involved with military procurement. In some cases the military may need to relax unnecessarily stringent specifications for ruggedness and security.

Even though smartphones can sometimes be used "as is," the direct adoption of commercial off-the-shelf (COTS) capabilities without modification is not always a viable practice for use at the tactical edge. Rather, modification of COTS products by the original developers or by third-party vendors to meet military requirements is increasingly common. A desirable goal is to modify COTS devices in a modular fashion so that there is an ability to evolve with the market-driven commercial evolution of the device. In some cases, vendors are undertaking some of the adaptations, because they see a growing market for secure, ruggedized devices that can be used not only in the military domain, but also in public safety, disaster relief, humanitarian aid, and wilderness applications. Other adaptations are being pursued by military and intelligence establishments. In these cases, the essential properties of the devices are being preserved. For example, the ability to run third-party apps is retained by employment of the Android operating system. Commercial chipsets, such as cellular LTE chipsets, are also being used.[*]

In the United States, *Nett Warrior* is a fast-track program to bring command and control network capabilities to the foot soldiers on the ground.[†] It is a basic, ruggedized smartphone that is mounted to a soldier's wrist, chest, or arm. The device plugs into the existing AN/PRC-154 Rifleman Radio, one of the radios of the JTRS that has already been fielded, in order to communicate over the tactical network. The proposed system includes the ability to project battlefield maps and unit location data to the user. The concept is not to deliver these to every soldier but to leaders of four-person teams. One early advantage of the Nett Warrior program is that the newer smartphone device replaces the original concept of having a backpack computer with a small display that flipped down over one eye and weighed around 14 pounds.[‡] The ability of the device to integrate with the JTRS radio should greatly increase its viability. However, early reports from field tests indicate that it is not yet ready for actual use, primarily due to issues with the capability to locate neighboring friendly soldiers who also have Nett Warrior.[§]

In 2012, the U.S. Marine Corps initiated the *Trusted Handheld Platform* program aimed at adapting commercial mobile device technology for secure communications. The goal of the project is to field commercial smartphones with standard hardware and software capable of accessing the military's classified and unclassified networks. The phones will be able to send and receive secure voice, data, and video across security domains. Several key commercially based technologies will be incorporated in the devices including virtualization and isolation methods including domain separation, process isolation, and resource encapsulation. Additional features that are desired include hardware root of trust, trusted boot, and Suite B encryption

---

[*] Agre et al. (2013).
[†] Stoker (2012).
[‡] Freedburg (2012).
[§] Munoz (2011).

meeting FIPS 140-2 certification from the National Security Agency. The solutions must be designed in a modular fashion to avoid reengineering of the commercial devices. The project involves collaboration between government and industry to speed up the certification process and also to result in device capabilities that can be adopted into future commercial versions. "The military and the commercial market share a common need—a highly secure, low-cost mobile device solution to share and manage sensitive content across their networks," according to Thomas Harvey, Senior Vice President, AT&T Government Solutions.[*] AT&T, one of several contractors, will provide 450 prototype devices based on the Android OS, but eventually the trusted platforms should support other devices and operating systems.[†,‡,§]

The U.S. Government also has a pilot project to provide secure communications over commercial cellular networks using commercially available Android-based smartphones. The approach is to provide a voice service using Voice Over IP encrypted with a second layer of software encryption in addition to the encryption provided by the vendor.[¶] In the future, the effort will expand to provide data capabilities. The phones will allow two users with the devices to have a secure classified conversation over the commercial cellular network. Some issues that the project is still working on include what to do about over-the-air updates and how to shift from inside a secure facility to an insecure network while maintaining the connection. The long-term goal is to provide these capabilities to the warfighter.

4.4.5 Implication 5: Military Establishments Must Still Do Specialized Research When Commercial ICTs Cannot Do the Job

Must the military resign itself only to adopting and adapting commercial technology, like everyone else? In the case of mobile networking and communications, the situation is not so simple. Military tactical deployment environments for mobile communications technology are different enough from commercial environments that many

---

[*] AT&T (2012).
[†] Kenyon (2012).
[‡] USMC (2011).
[§] AT&T (2012).
[¶] Iannotta (2012).

challenges critical to the military are unlikely to be solved by research and development in the commercial sector. Military establishments can still gain advantage by conducting and sponsoring their own focused research in communications and networking.

4.4.5.1 Lack of Infrastructure   The commercial sector's use of wireless technology, whether for cellular systems or enterprise wireless networks, is heavily based on well-considered deployment and maintenance of a supporting infrastructure. While the radio technologies are applicable, the infrastructure-dependent networking approach taken by the commercial sector is not so practical or robust for use in the tactical edge network domain, where infrastructure itself may be "on the move"—or at the very least undergoing various types of continuous disruption and dynamics.

Military network communications at the edge often involve more real-time collaboration and data sharing among groups of associated users and devices than in typical commercial applications, where centralization is more typically used to manage group network communications. There are good reasons for avoiding such strict central control and infrastructure in military systems. By doing so, one reduces the prevalence of "single points of failure" and also removes the necessity of installing infrastructure before starting an operation or doing "on the move" portions of a mission. It is also important to note that the challenges are not limited just to the formation and maintenance of dynamic, wireless network connectivity, but that they extend to all layers of the protocol stack.

The assumption of infrastructure-centric networking has resulted in many mature information services technologies that are dependent upon the somewhat centralized client-server paradigm, leading to more centralized points of failure even if the infrastructure enables more dynamic operation. Also, the relatively high capacity provisioning of infrastructure networks has allowed many of the modern information services to be developed at a rapid pace without a large amount of scientific understanding of the trade-offs that must occur for these to operate effectively and with high assurance in more dynamic or limited communication environments. More distributed models of network communication will probably need to be utilized to provide robust, effective tactical edge information services.

Distributed systems, particularly dynamic ones, are complex and difficult to analyze.

4.4.5.2 Multihop Networks   *Ad hoc* civilian wireless networks are typically "one-hop" wireless networks. That is, there is one wireless link from a user device, such as a cell phone or laptop computer, to a hub, and that hub is connected to other hubs or networks via wires or optical fibers. In military operations, however, there may be multiple wireless links for extended range or diversity, concatenated together before reaching a wired or fiber infrastructure. That is, many tactical military networks are multihop networks.[*]

Underlying some of the success of commercial mobile communications technology is the ability to accurately design and develop appropriate protocols and architectures and then predict performance prior to deployment. In general, the ability to deliver communications capability over one link is relatively mature. For single pair-wise communications links, there is a well-established "Shannon" theory,[†] and wireless channel models that guide engineers. In most cases the performance of a one-hop communication system can be predicted accurately enough to inform the design process. Unfortunately, multihop ad hoc wireless networks do not enjoy as strong a theoretical foundation.[‡] The tools used for early performance prediction of collections of many connected links (i.e., multihop networks) are often not of high enough fidelity to ensure that networking performance surprises are not discovered as the devices are built and tested in a variety of field environments. This makes it difficult to achieve trial-and-error cycles early in the engineering process.

4.4.5.3 Multiple Heterogeneous Networks   Commercial   networks have tended to be homogeneous. They have tended to solve problems within their boundaries and have often had strict control of the equipment in the network. Homogeneity greatly simplifies many networking challenges. Recently, the commercial world has begun to consider explicitly heterogeneous networks involving coexistence

---

[*] Braun et al. (2009).
[†] Shannon (1949).
[‡] Andrews et al. (2008).

of a range of different radio access technologies and WiFi, as well as cells of varying sizes.[*] However, the military is planning much more complex heterogeneous assemblages of multihop networks with connectivity on ground, in air, and in space. Making this work will require significant focused research in protocols, network management, and other areas.[†]

4.4.5.4 Complex and Contested Electromagnetic Environments   A reality of communications and networking for military operations is that some level of service must be preserved even when factors such as difficult terrain and lack of infrastructure make it almost impossible. Some level of service must also be preserved when adversaries are trying to disrupt connectivity. Chapter 8 discusses some of the effects of degraded communications on the C2 Approach.

In addition to the above problems, the proliferation of computers, sensors, and radios has resulted in a much more complex spectral environment for friendly forces. Finally, the hostile attacks and proliferation of emitters is further compounded by recent trends where military establishments are losing spectrum previously reserved for defense use. Spectrum is a precious resource, with recent auctions in the United States, Germany, and France indicating a commercial value, in the 700 to 800 MHz band, on the order of $1 per megahertz per person living in the country of consideration.[‡] Recent and projected spectrum losses by the U.S. Department of Defense may add up to hundreds of megahertz. In the hostile and complicated spectral environment, the spectrum must be used carefully and managed properly.

Government policy will help define some techniques for spectral relief, but there are also a number of technologies and architectural design approaches that will help make sure that military forces capitalize on any available spectrum. The first step will be to employ known techniques to make sure every transmission efficiently uses the available spectrum. This will result in the evolution or retiring of some more dated systems. New dynamic spectrum access technologies will

---

[*] Nokia Siemens (2011).
[†] Vassiliou et al. (2013).
[‡] Nachira and Mazzini (2011).

need to be further matured so that frequency-agile systems fill the gaps around existing systems. Significant research is ongoing in this area, but more is needed to fully develop broadband analog components and to develop frequency-band switching algorithms that are not susceptible to unsophisticated adversary attacks.

4.4.6 Implication 6: By Being Creative and Adopting Some Commercial Innovation Practices, Military Establishments Can Duplicate Some Level of Commercial Success

A number of successful research and development programs related to U.S. military command and control (C2), including the Tactical Ground Reporting (TIGR) system, the Command Post of the Future (CPOF), and the Combined Information Data Network Exchange (CIDNE) have displayed some common characteristics in their development processes. Among these are strong iterative links between end users and personnel in research, development, and engineering, and a relatively high degree of innovation by end users.[*] The development of these technologies can be characterized using a Kline "Chain Linked" model of innovation,[†] developed in the context of commercial and industrial product development. The model recognizes complex feedback loops between marketing and various stages of engineering and R&D. The technologies were also characterized by relatively rapid deployment to fill pressing user needs, and were often initially deployed "at risk," temporarily bypassing the normal U.S. military acquisition process. We do not go into more detail here; the reader is referred to Vassiliou et al. (2011).

4.5 General Remarks

The rapid advance and diffusion of advanced ICT has transformed the world, and it is transforming C2. Military establishments no longer have a monopoly on advanced technology of this type, and they no longer have as decisive an edge as they once did in processing and utilizing information for strategic advantage.

The Robustly Networked Environment opens the possibility of new enterprise designs and approaches for military establishments,

---

[*] Vassiliou et al. (2011).
[†] Kline (1985); Kline and Rosenberg (1986).

as it does for other establishments. Some of these approaches, with broader allocation of decision rights, broader patterns of interaction, and broader information distribution, as discussed in Chapter 6, may be more nimble and better able to confront the reality of Big Problems. While the Robustly Networked Environment may enable such approaches, the approaches are also, of necessity, dependent on that environment. Communications and information system capabilities—and attacks that degrade them—can thus limit which Enterprise Approaches are feasible or appropriate. The more decentralized, net-enabled approaches allowed by advanced technology may, in some cases, paradoxically be more resilient to attacks on that same technology. We discuss some of these issues in more detail in Chapter 8.

5

MEGATREND 3: UBIQUITOUS DATA

5.1 Introduction

A prevailing and instinctual impulse to achieve better situational awareness and a common operating picture is feeding the desire for more data in Command and Control (C2) operations. As unmanned vehicles and aircraft become more prevalent, the amount of data produced continually increases, posing new problems in processing, transmission, and analysis. Military establishments are continually fielding enhanced communications and computing infrastructure to obtain information from distributed C2 centers, as well as to supply and process the data from ever more sensor sources in real time. As this tendency is coupled with the widespread availability of the World Wide Web and other networked data sources, the inevitable result is that data is becoming ubiquitous.*

Net-centric concepts of operation require data at levels far exceeding anything imaginable in previous generations. The data demands of net-centricity influence, and are influenced by, the major megatrends identified in the other chapters in this book. "Ubiquitous Data" refers to the various aspects of this demand including collection, processing, storage, and distribution. That is, current systems can capture and store much more data, allowing many more dispersed people to have timely access to current as well as massive amounts of historical data. In this chapter, we examine the role of Ubiquitous Data and its impact on C2.

---

* We join many other authors, and newspapers such as the *New York Times*, in often using the word *data* as a singular, depending on context. We apologize to linguistic purists.

Information has, until very recently, been a scarce commodity. With the advent of the Information Age and the continued development of information and communications technologies (ICTs), individuals and organizations have had to manage a transition from a world characterized by the paucity of information to one marked by a deluge of information. The most successful enterprises are often those that can make this adjustment. Historically, organizations have been designed and processes put in place for vertical information flows, to funnel this valuable commodity to those in leadership positions. In the current data-rich environment, horizontal information flows are becoming increasingly important, and organizations are often adjusting accordingly to more decentralized and net-enabled structures, as we discuss in Chapter 6.

As information has become more widely available and as our ability to remotely access and disseminate information has increased, so has our interest in harnessing the power of information more effectively. As networks have replaced point-to-point communications and collaborative environments have been introduced, the proposition that information should be widely disseminated to promote shared awareness and self-synchronization is becoming prevalent. The theory of Network Centric Warfare (NCW) or Network Enabled Capability (NEC), introduced at the close of the last century,[*] has influenced the doctrines of militaries around the world. Military organizations recognize that any piece of information may have value to anyone in the organization. Since the scope of individual responsibilities, expertise, and experience will differ, people's "requirements" for information—and the quality of the information they require—will also differ. Efficient methods to handle these differences in data needs are needed.

5.2 Sources of Ubiquitous Data

The availability and demand for information have been rising at an extraordinary pace. We discussed the explosion in ICTs in Chapter 4, and presented statistics there that we will not repeat here. Given the proliferation of devices, degree of connectivity, and capacity to

---

[*] Cebrowski and Gartska (1998); Alberts et al. (1999).

communicate large volumes of data that we noted in Chapter 4, the makings of a data deluge are understandable.

Looking more closely at C2 systems, some modern sources of Ubiquitous Data causing the data deluge include the following:

- Growing sets of networked sensors and drones providing larger and more diverse sets of data than ever before
- More data sources for background information, such as the World Wide Web, social media network, or other online data repositories that can be accessed in moments
- The ability to store large quantities of historical data in low-cost, easily accessible repositories
- Multiple modes of communication such as e-mail, chat, voice, data stream, and video

Together these data sources need to be fused and analyzed, thereby enabling desirable capabilities such as enhanced situation awareness, both locally and globally. Global communications connect situational-aware C2 centers so that development and propagation of a common operating picture is possible in near-real-time that supports distributed operations. Locally, warfighters are often able to have live video feeds and other intelligence data concerning their immediate battlespace conditions, helping to make better real-time decisions on possible targets and to limit or prevent collateral damage.

The growth in military data production has been extraordinary over the last several decades. For example, the amount of data resulting from drones and intelligence, surveillance, and reconnaissance (ISR) technologies is estimated to have risen 1600% since the attacks of September 11, 2001.[*]

5.2.1 Sensors

Sensors have been increasing in numbers and evolving in capability for the last several decades. They can produce images with more pixels than ever before. They can operate in more energy bands, and multispectral sensors can operate in multiple bands. Sensors are also decreasing in

---

[*] Shanker and Richtel (2011).

size, weight, and power (SWAP), and also in many cases are available at lower costs. The SWAP reductions are enabling them to be carried on smaller vehicles (drones, aircraft, etc.), and lower costs are allowing emplacement in great numbers in various environments, such as the battlefield or the seafloor. The sensors are often coupled to a communications system so that their results can be transmitted to a control station or quickly downloaded when a carrier vehicle is returned.

There are a great variety of sensors producing different types of data including the following:

- Signals intelligence (collection of electronic intercepts or emissions)
- Moving target indicator data (tracks)
- Imagery (e.g., video, radar, laser, infrared, acoustic, electromagnetic) to identify objects
- Atmospheric, seismic
- Cyber intelligence (collection of network and computing intercepts and data)

The data from these sensors goes through processing, exploitation, and dissemination (PED) in order to produce meaningful information. Once information is extracted from the sensor data, the information needs to be fused with other relevant information to provide situation awareness. Each of these steps requires algorithms that are run on computing equipment. As the computing equipment—part of the Robustly Networked Environment—becomes more powerful and the algorithms are improved, the PED and information fusion steps can be shortened, providing enhanced information more quickly.

5.3 Problems Caused by Ubiquitous Data

While there are clear potential benefits associated with Ubiquitous Data, there are also serious potential problems if we cannot properly manage the resulting information flows. Some examples of the problems introduced by Ubiquitous Data include the following:

- *Operator Overload*—C2 personnel cannot cope with all the incoming information affecting their job

- *Data Volume*—The amount of data that needs to be communicated, processed, and stored is too great for the computing and communications systems to analyze in a timely manner and impact the mission
- *Data Quality*—The data is not trusted enough to be useful (e.g., it is not accurate enough, noisy, irrelevant, of unknown provenance, etc.)
- *Data Security*—The rules and procedures required to protect the data limit its utility or require too many resources
- *Data Sharing*—The rules and restrictions on the ability to share the data within an organization or between coalition partners limit its timely and effective use; also, systems may be incompatible, and the meaning of the data may not be mutually understood

As the amount and variety of data increase, an organization may be stressed at many levels to put the data to efficient use. There are both organizational and technical challenges that must be addressed, and there are many possible approaches to handling Ubiquitous Data. Several considerations from the organizational point of view include

- What are the effects of too much data on an organization? Are certain organization types better suited to handle large amounts of data?
- What can be done to improve data quality? How can we improve trust in the data?
- What are the effects of bad/erroneous/irrelevant data on an organization? Are certain organization types better suited to handle poor-quality data?
- Similarly, what are the effects of high-quality information on an organization? Are certain organization types better suited to handle high-quality information?
- Are there technological solutions that can be brought to bear to help address some of the organizational challenges? Can technology be brought to bear to effectively convert large amounts of data into high-quality information? How will organizations incorporate these technical changes in the future?

We discuss some of the above issues in more detail in the sections below. We also explore some of the questions immediately above in Chapter 8, using experimental evidence.

5.4 Operator Overload

There is mounting evidence that data overload is causing mistakes and errors in operational environments, particularly when the participants are under intense time pressure. For example, pilots often complain about all the information that is displayed on their cockpit screens.

In one reported case, it was found that drone operators failed to pass on crucial information that could have prevented a helicopter attack on Afghan civilians that killed 23 civilians, one of the worst civilian casualty incidents of the U.S. war in Afghanistan.[*] The drone operators were monitoring the drone video feeds and were involved with dozens of instant messages and message exchanges with intelligence analysts and troops on the ground. Despite reports that showed there were children present in the crowds, the drone operators did not focus on that fact and instead determined that there was a potential threat in the gathered crowd.

The environment of the drone operators is intense: "Every day across the Air Force's $5 billion global surveillance network, cubicle warriors review 1000 hours of video and 1000 high-altitude spy photos and hundreds of hours of 'signals intelligence'—usually from cell phone calls."[†] Their working conditions are also conducive to operator overload. There are often 12-hour shifts examining live video from drones and having multiple chats and phone conversations with headquarters, troops on the ground, and pilots in the area.

5.5 Data Volume

Even now, in some surveillance applications, data is being generated at a faster rate than can be processed, and it ends up being archived for later examination. With the increasing use of unmanned platforms, such as unmanned aerial vehicles (UAVs), the demand for information

---

[*] Shanker and Richtel (2011).
[†] Ibid.

delivered in real time to the tactical edge is also growing. Consider the Air Force Reaper drone-mounted "Gorgon Stare," which can transmit up to 65 video images per second,[*] or future systems such as the Defense Advanced Research Projects Agency (DARPA) Autonomous Real-Time Ground Ubiquitous Surveillance System (ARGUS-IS) platform with 1.8 giga-pixel video sensors generating data at 27 gigabits per second.[†] Such systems can quickly overwhelm the ability of C2 systems to process the information. As a result of this increase, many (ISR) decision support systems are receiving large volumes of data with poor control of data quality (e.g., with respect to noise and clutter). Requests requiring special analysis methods and unpredictable data requirements are normal occurrences. Tracking vehicles in an urban environment or identifying roadside bombs from video are typical examples of particularly challenging requests. There is also the problem of short- and long-term storage, data accessibility, and supply of the computational power to process the requests. In this context, timeliness becomes a critical property. If the raw or processed data is not available to track a target, then it quickly becomes of limited value.

5.6 Trustworthiness of Data

For C2 systems, determining the level of trust to place in data can be extremely important. It is often difficult to determine whether separate reports are referring to the same or different incidents. Detecting data ringing, where the same report is relayed by different individuals, can be a serious challenge. Similarly, "copy and paste" is frequently used in report generation, and without automated tracking of sources from copy-and-paste operations, it is difficult to assess trust. Incorporating some form of provenance information may help clarify these situations. Within the U.S. Department of Defense (DoD), the various armed services are currently focused on defining Authoritative Data Sources (ADSs) and using a standardized metadata registry for data discovery and use. These systems have limited provenance information, primarily containing only the source and date. Outside of the authoritative sources, there is almost no provenance tracking.

---

[*] Kenyon (2011).
[†] Robinson (2009).

For certain applications, such as bioinformatics or physics, it is appropriate to capture the entire workflow that transformed the data from input to output for purposes of validation or repeatability.[*][†] Other applications mainly require documentation of original sources, context, or other relevant pieces of information. In theory, provenance information should allow one to determine the complete history of the data in question. For example, a "W7" model of provenance (What, Who, When, Where, Which, How, and Why) has been proposed that captures all relevant information for full documentation of a data life cycle from creation to destruction. However, this can require huge amounts of storage in practical scenarios.[‡] For resource-limited environments, such as those often faced by C2 systems, there are limits to the amount of provenance information that can be collected, stored, or transmitted.

5.7 Sharing and Interoperability

As we shall see in Chapter 7, interoperability problems and an inability to share data are major causes of C2 failure.

Interoperability problems are caused by both policy and technical issues. Each of the U.S. Armed Services maintains its own family of C2 systems that are tailored to their particular mission needs: air, ground, sea, space, and special operations. For example, the Global Command and Control System (GCCS), a family of C2 systems, includes over 200 systems or services and is intended to have worldwide reach and incorporate components from all service branches.[§] Data must be exchanged among the systems in the same family as well as with other nonfamily systems. In joint and coalition operations, each participating service or nation comes with its own C2 systems. U.S. joint commands employ C2 systems that must combine information from the multiple services. Coalition commands must exchange information among the services and with C2 systems from other countries. These information-sharing requirements bring up significant problems of how to properly

---

[*] Cohen et al. (2006).
[†] Moreau et al. (2008).
[‡] Ram and Liu (2008).
[§] Ceruti (2003).

control access to data, and often how to control data crossing security classification domains (multilevel and cross-domain security).

The problem of data exchange between C2 systems has been well known for many years in joint and coalition settings, and several key developments have been achieved, such as the NATO Network Enabled Capabilities (NNEC) Common Operating Picture (COP).* The NNEC COP addresses issues such as standards, dynamic tailoring, multilevel security, provenance, and knowledge management (timeliness and access).

There are various representations of the C2 domain that have been in use for some time. In particular, there is the Joint Consultation, Command, and Control Information Exchange Data Model (JC3IEDM) that is in use by many countries and also by NATO. JC3IEDM exchanges are not XML-based internationally, but JC3IEDM is the U.S. Army's chosen data model for information exchange.† Within the U.S. DoD, recent efforts have been attempted to define a common data model, such as the Universal Core (UCore) standard for information exchange between systems,‡ but these efforts have not been adopted. Efforts are ongoing to come up with a reasonable solution acceptable to the services, including adoption of variants of the National Information Exchange Model (NIEM),§ which has been successfully employed by law enforcement agencies.

5.8 Data and Information Quality

Many of the issues faced by C2 operations in handling data are similar to those faced by other enterprises. Frameworks for describing issues related to data are evolving in the business world and can be used to describe common issues related to many aspects of data. One such framework called Total Data Quality Management (TDQM)¶ is very useful as a basic template for categorizing data quality issues, both for specific applications and for large enterprise perspectives.

---

* Waters et al. (2009).
† U.S. Army (2005); NATO (2012).
‡ Green (2009).
§ USG (2014); Takai (2013).
¶ Pipino et al. (2002).

The transition to a net-centric environment and the increasing automation of C2 functions make the quality of the underlying data, upon
which decisions and actions are based, critical to success. Operating on
bad data can have serious consequences, especially in a military context.
In the commercial arena, it is estimated that operating on poor data has
an economic cost of about $600 billion annually.* A few of the many
side effects of poor data quality include delays associated with reconcil-
ing data, loss of credibility, customer dissatisfaction, compliance problems, delivery delays, lost revenue, and loss of trust in automation and
computing systems. Properties that reflect good data—accuracy, integrity, provenance, and timeliness—as well as the ability to share the data
with others and to have a common understanding of its meaning are
intuitively desirable but are not routinely incorporated in today's complex systems. In part, this is because the underlying architectures do
not make data quality a primary objective of system design. In the military C2 domain, the effects of poor data can have even more disastrous
consequences than in other domains (see Section 5.12 and Chapter 7).
Making quality considerations an inherent part of the design and maintenance processes of C2 systems should benefit the decision making.

Data is a resource that must be managed, protected, and preserved
across its life cycle like any other. The dominant issues confronting
data management in large enterprises have been frequently reported
and include missing or incorrect data, missing or incorrect metadata,
redundant data storage, varying data semantics, and nonstandard data
formats. Data portability (freeing the data from stove-piped applications) is also a common concern in both military and commercial
domains. These issues are of prime concern in C2 systems.

For C2 functions, data is used to develop situational awareness and
a common operating picture by which commanders make decisions
and effect control. Commanders require many types of data, ranging
from logistics to weather to geospatial to tactical information, to support various warfighter operations. Data must be collected, analyzed,
and communicated via various manual and automated messages, and
exchanged between various C2 systems and people. A commander
has little control over the sources that supply data to the C2 systems,
especially in times of crisis. Each C2 system may store portions of

---

* Sebastian (2008).

current data and maintain some amount of past data for historical analysis purposes. The tempo of activity and the volume of data on which the system depends are both rapidly increasing, revealing many stress points in the current systems. In general terms, a modern C2 system is a large, heterogeneous distributed, real-time processing system that is resource limited (bandwidth and computation power) at some of the end-points, with frequent disruptions and highly dynamic information flows. The data is segregated among multiple classification levels and is contained in multiple, distributed storage facilities and heterogeneous databases. As data is delivered with higher frequency and larger volume from more places, decision makers must become more responsive.

Data and information quality can be defined and their properties measured, much like other quality-controlled entities or processes. The terms "information" and "data" are often used interchangeably, by nearly everyone, including us in this book. There is, however, a distinction that needs to be made for the purposes of the discussion in this particular section[*]:

- *Information* is defined, for the purposes of this section, as knowledge concerning objects, such as facts, events, things, processes, or ideas, including concepts, that within a certain context has a particular meaning.
- *Data* is defined, for the purposes of this section, as the reinterpretable representation of information in a formalized manner suitable for communication, interpretation, or processing.

In the context of this chapter, data includes both raw and processed information and "all data assets such as system files, databases, documents, official electronic records, images, audio files, Web sites, and data access services."[†]

Data quality can be simply defined as the fitness for use of the data.[‡] A more practical definition is the degree to which data "meets the requirements of its authors, users, and administrators".[§] The key

---

[*] ISO/IEC (1993).
[†] Simon (2006), p. 21.
[‡] Strong et al. (2007).
[§] IAIDQ (2014).

point to be taken from these definitions is that the generic notion of the quality of data, like many other ideals of quality, is dependent on context, or intended use. Data quality metrics can be absolute or fitness-for-use (relative) measures, while information quality, because information has meaning only in a context, is primarily a fitness-for-use measure because it takes into consideration the needs of the consumer of information and the circumstances surrounding its use. While a given piece of information has a unique quality value associated with each of its absolute attributes, there can be different "fitness" scores for the same piece of information, and these can differ widely as a function of the consumer, the task at hand, and the prevailing circumstances. For example, some consumers need information with a high degree of precision while others can tolerate a degree of imprecision. Hence, a particular level of precision may receive an unacceptably low score from one consumer and a high score from another. In fact, a piece of information can be highly valued by a consumer at one point in time and have no value to the same consumer at another point in time.

Information quality is closely tied to data quality and the terms will be used interchangeably unless specifically noted. Given that virtually all information that individuals receive and share comes in the form of data (the exception is direct observation or retrievals from memory of the brain), it is important to address the quality of data as it provides a common denominator across enterprises, missions, and circumstances.

Given that data is such a pervasive part of any information technology (IT) system, many ways of partitioning its quality properties have been suggested. In some early data quality research, data was primarily characterized by accuracy, completeness, timeliness, and standards (ACTS). This basic list has been expanded over the years in many directions by different communities. In particular, since the early 1990s, a Total Data Quality Management (TDQM) research community has expanded ACTS to 16 data quality dimensions grouped into 4 broad categories and successfully used them in assessments of an organization's data quality environment.[*]

The intrinsic properties (also known as inherent or absolute) relate to the accuracy and pedigree of the data and do not change

---

[*] Pipino et al. (2002).

depending on environment or intended use. Accessibility, in this usage, refers to the system properties such as how and where the data is stored and the means of protecting the data, such as access control. Contextual properties depend on the application for which the data is used and can have temporal behavior. The representational properties are the more common notions of standardization and interoperability. These categories help to show how the characteristics are related to each other and the environment in which they are situated.

The dimensions are defined more precisely as follows:

- Intrinsic properties
  - Freedom from Error—The extent to which data is correct and reliable.
  - Believability—The extent to which data is regarded as true and credible.
  - Reputation—The extent to which information is highly regarded in terms of its source or content.
  - Objectivity—The extent to which data is unbiased, unprejudiced, and impartial.

- Operational properties
  - Accessibility—The extent to which data is available, or easily and quickly retrievable.
  - Security—The extent to which access to data is restricted appropriately to maintain its security.

- Contextual properties (or fitness-for-use)
  - Relevance—The extent to which data is applicable and helpful for the task at hand.
  - Timeliness—The extent to which data is sufficiently up to date for the task at hand.
  - Completeness—The extent to which information is not missing and is of sufficient breadth and depth for the task at hand.
  - Amount of Information—The extent to which the volume of data is appropriate for the task at hand.
  - Value Added—The extent to which data is beneficial and provides advantages from its use.

- Representational properties
  - Conciseness—The extent to which data is compactly represented.
  - Consistent Representation—The extent to which the data is presented in the same format.
  - Ease of Operations—The extent to which data is easy to operate on and apply to different tasks.
  - Interpretability—The extent to which data is in appropriate languages, symbols, and units and the definitions are clear.
  - Understandability—The extent to which data is easily comprehended.

All of the communities that have enumerated data or information quality metrics (e.g., product information for purchase of goods, enterprise IT, intelligence evaluation, software engineering, C2) have included both absolute and fitness-for-use metrics that correspond roughly to the above categories. Some metrics focus on the inherent accuracy and representational properties of the data, and others focus on the ability of a consumer to understand and use it for a particular set of applications.

C2 systems are generally beset with similar data quality issues as in the information technology (IT) enterprise community addressed by TDQM. However, several data-quality attributes are of relatively greater importance in C2 due to the potential lethality of errors in decision making. In particular, timeliness, accuracy, completeness, believability, and interoperability are critical to successful C2 decision making. These characteristics or dimensions primarily consider data and information from the perspective of an individual's role in an organization (e.g., database administrator, analyst, commander) rather than from the enterprise point of view. We will extend these notions to the enterprise level in a later section.

Data sharing and accessibility are areas that began to receive increased attention after the attacks of September 11, 2001. The intelligence community (IC), in particular, is also very worried about "spoofing," the injection of false data that can corrupt decisions or analyses. There is a great need for provenance information to track sources and the intermediate handling of data to detect deliberate deception attempts. Another concern is that of inconsistent data that can arise from multiple observers. Nonauthoritative sources of data are also a persistent problem, and proper weighting is needed. In some C2 systems, such as the Global Command and Control System (GCCS), the data is generally vetted and considered authoritative, while in others, such as the Tactical Ground Reporting System (TIGR), the data can be entered by any user who observes an interesting event. Both types of systems have their uses, but the differences show that the pedigree of data should be an explicit factor. Another interesting IC and C2 issue is that information that was presented as true may later be found to be untrue, and that this meta-information needs to be disseminated as well. Some data quality properties, such as timeliness and accuracy, can have a more severe impact in a C2 tactical situation. It is not acceptable, for example, to target the wrong building due to incorrect data.

The U.S. Department of Defense has recognized data quality as an important issue in the last decade and incorporated aspects into its overall data management strategies such as the DoD Net Centric Data Strategy (NCDS)* and the Army Data Transformation (ADT).† The NCDS claims that its goals do not include data quality or accuracy considerations, but that achieving the goals should result in improved data quality and accuracy. The ADT plan is aimed more specifically at processes to improve data quality as the systems are transformed to net-centric operations. The ADT has indicated six phases in which it is working to improve data management and data quality:

1. Accountable—Incorporate common data standards and governance practices.
2. Authoritative—Identify and manage master data elements and authoritative sources.
3. Transform—Employ standardized structures and schemas such as data yellow pages to improve data sharing.
4. Expose—Make data accessible and responsive to users through the Army Data Services Layer (ADSL). Four methods of exposing data are messaging, data services, data warehouses, and data security.

---

* Walsh et al. (2009).
† U.S. Army (2010).

5. Register—Validate data schemas and services against standards and then register in repositories (e.g., authoritative data repository) to enable visibility and reuse.
6. Assess—Monitor and assess data maturity levels using metrics. Measure the progress in improving data quality.

In the C2 domain, the C2 research literature has coalesced around a multidimensional definition that encompasses both absolute and fitness-for-use metrics.[*] The age or currency of information is an example of an intrinsic measure while the timeliness of the information (relative to when it is needed to make a decision) is an example of a "fitness-for-use" measure. More recently, as the awareness of information and communication-related threats has increased, measures of information security have been added to the list of fitness-for-use measures. Quality-of-Information attributes for C2 systems involve three sets of variables, as shown below.[†] There are interdependencies among these measures. For example, the levels of trust and confidence are related to the values of one or more of the absolute measures, and the measures of information security. Thus, information quality needs to be considered holistically:

- Intrinsic (Absolute)
  - Correctness—The extent to which information is consistent with ground truth.
  - Consistency—The extent to which a body of information is internally consistent.
  - Currency—The age of information.
  - Precision—The degree of refinement, level of granularity, or extent of detail.

- Contextual (Fitness for Use)
  - Relevance—The proportion of information that is related to the task at hand.
  - Completeness—The percentage of relevant information attained.
  - Accuracy—The degree of specificity relative to need.

---

[*] Alberts et al. (2001); NATO (2006).

[†] Definitions for the absolute and fitness-for-use measures are taken from Alberts and Hayes (2006). Definitions of *information security measures* taken from OSPA (2013).

- Timeliness—The availability of information relative to the time it is needed.
- Trust—The credibility of the information source.
- Confidence—The willingness to use the information.

- Operational (Information Security)
  - Integrity—Unchanged from its source and has not been accidentally or intentionally modified, altered, or destroyed.
  - Availability—The assurance that data transmissions, computer processing systems, and/or communications are not denied to those who are authorized to use them.
  - Authenticity—Having an undisputed identity or origin.
  - Nonrepudiation—Assurance that the sender of data is provided with proof of delivery and the recipient is provided with proof of the sender's identity, so that neither can later deny having processed the data.
  - Confidentiality—Accessible to only authorized users.

In Table 5.1, we present an initial comparison mapping from the data quality concepts of various domains: the 16 TDMQ dimensions of IT systems, a combination of the NCDS goals (and the ADT phases) from the military domain, the intelligence community perspective,[*] the ISO 8000[†] for commercial product descriptions, ISO/IEC 25012 for software product quality,[‡] and the C2 Research Domain.

In the NCDS column of Table 5.1, we indicate in parentheses the phases of the ADT expected to have the most impact on data quality. The NCDS fails to address certain properties, particularly timeliness, which are critical to C2. Also, although the table has indicated that NCDS covers some areas such as believability and reputation, the extent of this coverage, which is primarily limited to using authoritative data sources that have been vetted, does not span many of the situations frequently encountered in C2, such as data from a variety of sources with varying pedigree (provenance, reliability, etc). The NCDS notion of assessment is not well captured in TDQM or the C2 Research Domain, but it is an important factor in maintaining data quality.

---

[*] Zhu and Wang (2010).
[†] ISO (2009).
[‡] ISO/IEC (2008).

Table 5.1 Comparison of Data Quality Metrics from Differing Application Areas

TDQM	DOD NCDS DATA GOALS	INTELLIGENCE COMMUNITY	ISO 8000	ISO 25012	C2 RESEARCH DOMAIN
INTRINSIC					
Free of error	Trusted	Accuracy	Accuracy	Accuracy, precision	Correctness, precision
Reputation	Trusted (accountable, authoritative)		Certification		
Believability	(accountable, authoritative)		Certification	Credibility	
Objectivity (provenance)	Trusted (accountable, authoritative)	Objectivity	Provenance	Traceability	
					Currency
OPERATIONAL (ACCESSIBILITY)					
Accessibility	Visible, accessible (expose)	Usability		Accessibility, availability, portability, recoverability performance	Availability
Security (access control)	Trusted (expose)			Confidentiality	Integrity, confidentiality, authenticity, nonrepudiation

CONTEXTUAL					
Amount of data					
Relevance	Responsive to users' needs	Relevance			Relevance
Value added		Readiness			
Timeliness	Responsive	Timeliness		Currentness	Timeliness
Completeness			Completeness	Completeness	Completeness
					Accuracy, trust
REPRESENTATIONAL					
Understandability	Understandable	Usability	Master data open tech. dict.	Understandability	
Conciseness					
Ease of operation (manipulation)				Performance	
Interpretability	Interoperable	Usability	Master data: syntax		
Consistent representations	Institutionalized, interoperable (transform, register)		Master data: conformance	Consistency, compliance	Consistency

Source: Agre et al. (2011).

For the intelligence community, it appears that usability covers several areas that are subjective and would be difficult to measure objectively. Also, interestingly, the other domains do not seem to capture the IC notion of readiness, which indicates that the data should be adaptable to changing circumstances and requirements.

The ISO 8000 and related standards provide a broad range of coverage. However, they do not adequately address the broad operational and contextual dimensions, including some important issues, such as timeliness or ease of operation. In terms of long-term impact, though, the ISO 8000 may well have the most influence on future organizations due to its widespread adoption in the automated purchasing arena.

The ISO 25012 specification for software is of particular interest to current and future highly computerized C2 operations. There are several operational characteristics related to security that are relevant to C2 systems and not captured by TDQM. In particular, the notion of precision is included with accuracy as an inherent attribute. In addition, the 25012 specification includes operational characteristics such as performance, availability, recoverability, and portability that are important in automated, data-intensive systems such as those employed in C2.

The C2 Research Domain characteristics have several features that are different from those of the other domains. In particular, currency is properly viewed as an inherent property of the data and timeliness viewed as the contextual aspect. The age of the data is an absolute fact, while its utility depends on circumstances. Similarly, the characteristics of trust and accuracy are considered as contextual features, unlike the TQDM viewpoint. For different situations, the accuracy required to make a decision may vary. Similarly, the amount of trust in the data may depend on the necessity to make a decision. We also note that the notions of representation of the data, while clearly important from a data-sharing perspective, have not been extensively explored in the C2 Domain to date.

5.9 Metrics and Tools for Data and Information Quality

It is useful to employ metrics to quantify the quality of the data under consideration and to make economic or strategic decisions on how to improve or maintain a given quality level. Researchers have proposed a variety of metrics that can generally be divided into objective and subjective measures, but their interpretation is typically context dependent. For instance, in some applications such as digital voice storage and transmission, it is acceptable to have a percentage of missing data without appreciably degrading the quality. In other applications, a missing value could be catastrophic.

The metrics for the 16 TDQM features are defined in terms of three basic forms.[*] These are (1) a simple ratio, (2) "min or max," and (3) weighted average. The metrics are typically normalized between 0 and 1. Using a simple ratio, it is possible to represent completeness, accuracy, precision, consistency, concise representation, relevancy, and ease of manipulation. For example, an accuracy metric can be a simple ratio of the number of accurate records divided by the total number of records. The criteria for acceptable accuracy are a function of the context or application. These metrics can be defined for high-level notions but may be made more specific to satisfy the circumstances, such as schema, column and population completeness in a database.

Min or max operations can be used for metrics that are composed of several underlying dimensions. Examples include believability, timeliness, accessibility, or amount of data. For example, timeliness has been defined[†] as

$$\max [0, 1 - (\text{age at delivery/shelf-life})]$$

where *age at delivery* is the delivery time minus the data creation time, and *shelf-life (volatility)* is the total length of time that the data is valid and usable. If the age is less than the shelf-life, then the data is still usable. The earlier the data is delivered, the more time there is to process the data, and thus, the larger is the metric. In other studies, other functional forms to represent the decay of timeliness are employed, and the function is often weighted by an exponent to magnify the effects of the timeliness.

The weighted average metrics are used if there is enough detailed information on the underlying features to determine their relative contributions. In addition, weighting the simple measures can allow incorporating notions of criticality, utility, and/or costs.

---

[*] Pipino et al. (2002).
[†] Ballou et al. (1998).

Some metrics are naturally objective and others subjective. "Believability," for example, is subjective and must be assessed from user opinion or surveys rather than direct measurements or observations. Metrics have been developed for each of the TDQM dimensions based on subjective and objective surveys of both users and system owners. The exact forms of the metrics or the weighting of the metrics depend on the various contextual situations. For example, timeliness may be more critical in some applications than in others. An interesting observation is that the subjective results often differ depending on the perspective of those interviewed.* For example, the believability of the data is often judged differently by the users and the data system owners. Discrepancies such as this indicate that further analysis may be necessary to discover the underlying data properties.

There are many tools available in the commercial and open-source domains to support data quality measurement and improvement. Data validation tools examine data as it is input into the system and reject or correct data item errors. Extract-transform load (ETL) tools can sometimes be configured to perform validation functions as the data is prepared for export and entered into an existing data set. Data profiling or data auditing tools examine a data set to identify problems such as missing, duplicate, inconsistent, and otherwise anomalous data, and also to compute data quality metrics. Data cleansing (or scrubbing) tools go through an existing data set and attempt to detect, correct, or remove troublesome data items (incorrect, incomplete, inaccurate, etc.). Many variations exist in the market, with some tools using complex reasoning and rules on relations to correct data sets. Data cleansing can be quite time consuming on large data sets and efficiency is a key consideration. Data monitoring tools are used to maintain the data quality over time as the data set is used.

It is well known that one-time attempts to improve data quality are not sufficient because data degrades over time due to factors such as data change, system change, and migration. For example, data on people can change rapidly due to change of residence, death, marriage,

---

* Pipino et al. (2002).

divorce, and so forth. Computer operating systems are updated and patched frequently to improve functionality, to fix bugs, or to respond to vulnerabilities. It is generally accepted that a continual process to monitor data quality is necessary if there is to be hope of maintaining or improving data quality. Also necessary are clearly defined policies and governance to ensure best practices are followed and to prevent accidental loss of quality.

Several methods have been proposed to help organizations manage data quality continuously in order to achieve desired levels. One popular method, based on a diagrammatic scheme called Information Production Maps (IPMs), models data as a product that goes through manufacturing stages similar to an actual physical product in a manufacturing plant and applies similar quality management procedures.* IPMs are particularly useful for dynamic decision environments such as an e-business, or C2 systems, where timely quality information can have a large impact on effective decision making.

5.10 Enterprise Information Quality

In addition to the absolute and fitness-for-use measures that pertain to various aspects of the quality of individual pieces of data and information (such as accuracy, timeliness, value, and usability) in the context of individual decision making, it is important to understand information-quality needs from an enterprise perspective. This involves knowledge about the total set of the information available, the nature of enterprise information flows or information-related transactions, how information is distributed among individuals and component organizations in an enterprise, how this impacts information quality, and ultimately how this affects enterprise decision making as determined by organization, policy, and process. Measures of enterprise information quality provide us with the ability to measure the larger set of information available, how the quality of this set evolves over time, and the information flows and transactions that determine the evolution in quality. These measures will enable us to explore several

---

* Shankaranarayan et al. (2003).

of the hypotheses of interest concerning organizational approaches. Some example enterprise measures of interest include:

- Distribution of Information—Who has access to what information.
- Extent of Shared Information—Information in common, a prerequisite for shared awareness.
- Responsiveness—Time for the enterprise to reach a decision.

Other measures capture enterprise-level views of the individual characteristics such as timeliness or completeness. At the enterprise level, due to the interdependencies among the quality attributes, many behaviors are emergent and not predictable based on the individual characteristics. For example, the amount of information shared can impact the system performance, creating delays in transporting information that impact enterprise timeliness.

Collectively, enterprise measures of information quality help to tell us the "information position" of the enterprise. In the literature of NCW, information superiority is determined by comparing the relative information positions of adversaries or those who share a competitive space. A central hypothesis is that, all things being equal, those that have information superiority will prevail. Huge investments in IT and information-related capabilities have been made based, in part, on this assertion.

Experimental results bearing on the relationship between *organizational approach* and *enterprise information quality* are described in detail in Chapter 8.

The next section explores some of the effects of information quality on actual C2 operations.

5.11 Information Quality and C2

From a C2 perspective, the key data issues that are frequently discussed include interoperability, distributed access, timeliness, accuracy, provenance, and security. There are also issues with information overload, as the volume of data that is available, both from the tactical and strategic sides, is rapidly increasing. The data needs to be processed in a timely manner, incorporated into the common operating picture, and delivered where needed. There are also issues associated with limited or disadvantaged communications capabilities (see Chapter 7). This limits data availability, and C2 systems must accommodate these resource-constrained situations. Looking at this from the data quality perspective, we see that most of these issues are covered by the data quality properties discussed previously. The ideal collection of characteristics for C2 systems is still an evolving set that is changing with the evolution of C2. A set of metrics, based on the above assessment of the strengths and weaknesses of the approaches from various perspectives, should be employed in evaluating C2 systems. However, a data quality strategy specifically for C2 should emphasize and tailor these dimensions.

5.12 Examples of Information Quality Effects on C2 Operations

In Chapter 7, we present many cases of C2 failures caused, at least in part, by the inability or failure to communicate essential information. Here, we will concentrate on a few examples specifically related to data quality. There are a number of cases of dramatic effects that are at least partially due to C2 data quality problems. The May 7, 1999, bombing of the Chinese Embassy in Belgrade (Serbia, former Yugoslavia) by U.S. planes, conventionally interpreted as unintentional, was caused by a systemic failure in the targeting process, but was also plagued by data issues.* One example was the use of older map data that failed to show the updated location of the embassy after a move in 1996. Also, the actual address of the intended target (a warehouse) was only estimated, and not carefully verified against a map with accurate address information. Other problems were caused by duplicate target requests that appeared to come from different sources but were ultimately from the same source. (This is sometimes called "ringing" and can be caused by a lack of provenance tracking.) Furthermore, there was a failure to check the target against a database of known off-limits targets.

Data quality issues have also been identified in two other disasters: the space shuttle *Challenger* explosion on January 28, 1996, and the

---

* Myers (2000).

shooting down of an Iranian Airbus by the *USS Vincennes* on July 3, 1988.* The Presidential Commission investigating the *Challenger* disaster cited flawed decision making surrounding the possible problem with O-rings at cold temperatures. The attack on the Iranian Airbus was also attributed to flawed decision making under time pressure, when the ship identified the passenger plane as a hostile military jet in attack mode. From the data quality perspective the decisions were affected by lapses in accuracy, completeness, consistency, relevance, and fitness-for-use in the *Challenger* case, and accuracy, completeness, consistency, fitness-for-use, and timeliness for the *USS Vincennes*. For the space shuttle *Challenger*, the data needed for proper analysis was available but not properly used and not presented in a form that assisted the management to make correct decisions. For the *Vincennes*, the initial misclassification occurred when users did not realize that the system reused a target designation number and then failed to resolve the resulting inconsistencies. Given all the pressures of decision making, it is arguable that data issues contributed to the erroneous decisions.

A case study of Operation *Anaconda* in which the U.S. Army successfully defeated Al Qaeda forces in the Shahikot Valley of eastern Afghanistan in March 2002 showed many problems that can be partially attributed to C2 data quality.†,‡ Though the operation ultimately succeeded, the initial battle plan required extensive modification. It was designed to last for a week; however, the battle lasted 17 days, and resistance was much stronger than anticipated, requiring much more air support. Some of the problems were related to the quality of the intelligence data, such as inaccurate and incomplete estimates of enemy forces and their willingness to fight, or the disposition of civilians. The intelligence data, which relied primarily on human intelligence, was not properly verified and vetted, reflecting believability and accuracy issues. The satellite imagery was often three days old.

---

* Fisher and Kingma (2001).
† Kugler (2007).
‡ MacPherson (2006).

There were also interoperability problems among and between joint and coalition forces.* Some of these interoperability issues arose from a lack of unity of command, due to the relative newness of the Army forces in the area, and lack of command authority over Special Forces, air support, and Afghan allied forces that were all part of the operation. Although communications reportedly worked for each U.S. service component, problems occurred in communicating with other services and with allied Afghan forces. For example, "Army personnel could use their FM radios to communicate directly with overhead Navy and Marine Corps aircraft but not USAF aircraft, such as F-15Es and bombers." Also, U.S. gunships mistakenly fired on an allied Afghan column, partially causing them to turn away from the area. In addition, long-range communications between headquarters and edge forces was bandwidth limited, and communication between headquarters and central command was inconsistent (timeliness, accessibility). Finally, there was a lack of common understanding about the differing rules of engagement and procedures governing Close Air Support (CAS), reflecting understandability problems.†

5.13 Technological Solutions

Current practices in C2 typically involve a human in the loop for almost all levels of data entry, analysis, and decision making. The increase in data volume is overwhelming both the people and the systems and causing mistakes induced by information overload. Increased use of automated machine processing of the raw data and elementary decision making appears to be necessary if modern commanders are to operate effectively under this data deluge. The commanders must be involved at the crucial decision points and provided with timely situation awareness, but otherwise not encumbered by the lower-level data details. Decision support systems that can process raw data and make the low-level decisions, alerting the commanders at the crucial points are needed. Incorporation of information quality

---

* We have already seen interoperability problems in connection with the 1980 U.S. hostage rescue attempt in Iran, discussed in Chapter 1. Chapter 7 will also discuss interoperability issues in other cases.

† Kugler et al. (2009), p. 20.

characteristics, both at the individual and enterprise levels, along with other forms of metadata that are semantically defined and can be processed and understood by the decision software and presented in a meaningful way, may go far in facilitating this environment.

There are several key emerging technological solutions that are being brought to bear on issues related to Ubiquitous Data:

- *Data Curation*—This involves automated processing of incoming data to create and associate appropriate metadata that reflects properties of the data such as provenance and accuracy.
- *Big Data Analysis*—These are techniques that can process an extremely large set of data and reduce it to valuable information.
- *Onboard Processing*—This involves equipping sensor platforms with computing and communications capabilities to process sensor data and extract the truly critical information, such as tracks. This critical information can then be communicated instead of the raw data, reducing demands on communications links and on warfighters.
- *Human-Computer Interface*—Improved methods to capture and present information to C2 operators and decision makers can reduce effects of data overload and present information in the most effective fashion.

5.14 Data Curation

Automation of data handling is one of the keys to overcoming the pressures of increasing data volume and reducing the severity of the data overload problem. Digital data curation refers to the methods and practices of preparing and managing data for use in computer-based analyses over the life cycle of the data. Data curation enables automated data discovery, advanced search and retrieval, improvment in the overall data quality, and increased data reuse.

As described earlier, many important and complex C2 activities require use of disparate data sources (structured and unstructured) that are time varying, at various levels of quality (completeness, accuracy, etc.), and of ambiguous origins. Currently, dealing with such disparate data is manually intensive and expensive, in large part due to problems with the quality of the data and its ability to be quickly processed.

Data curation methods can help to solve such problems by pre-positioning the data so that it can be quickly assembled to answer unanticipated questions, and maintaining the data in a form that facilitates automated processing by computer tools. Data curation can be applied to structured, semistructured, and unstructured types of data. Unstructured data is generally textual information, such as found in a technical report, a news article, and so on. Structured data is information that has been formatted and associated with additional information that defines its meaning. For example, data in a database has a schema, which describes fields with labels and well-defined semantic meanings, such as "Last Name," so that this information can be processed by computing algorithms. Semistructured data refers to data content that is labeled, but the data itself can be free-form text. For example, a response to a survey question may fall into this category. Clearly, structured data is more amenable to computer-based analyses. Techniques for working with unstructured or semi-structured data, enabled by Natural Language Processing (NLP), are increasingly becoming commercially available, so that analysis of textual data is becoming practical. Other emerging technologies such as machine learning, Big Data computational methods, and visualization are enabling advances in data curation. *Wolfram Alpha*[*] is an example of a commercial capability incorporating data curation techniques.

In general, the more structure that can be associated with data, the easier it is to perform analysis. The intent of data curation is to provide structure to the semi- and unstructured data so that automated analyses are possible with less manual effort. At the current stage of development, data curation is not a fully automatable procedure but can reduce the overall effort significantly, increasing the timeliness of analyses and helping to reduce the operator overload problem. There is a clear relationship between data curation and data quality, and data quality metrics can be used to gauge the effectiveness of data curation.

The data curation process is a sequence of steps that provides metadata that will improve the data quality and assist in further processing.

---

[*] Wolfram (2010); Wolfram|Alpha (2014).

The data curation process can be described using what we might call the "Seven C's" of data curation[*]:

1. *Collect*—Interface to the data sources and accept the inputs
2. *Characterize*—Capture available metadata
3. *Clean*—Identify and correct data quality issues
4. *Contextualize*—Provide context, provenance
5. *Categorize*—Fit within framework that defines the problem domain
6. *Correlate*—Find relationships among the various data
7. *Catalog*—Store and make data and metadata accessible with application program interfaces (APIs) for search and analysis

In a data curation process applied to an operational environment, the *collect* step involves automated procedures that capture the data, format it, and store it in the appropriate data repository, such as a relational database for structured data or a NoSQL database for textual documents. The data would be stored in common formats such as JSON or XML. *Characterization* is applied as data is captured, where additional metadata such as creation time, capture method, sensor settings, accuracy, precision, and so on, are also recorded along with the data. In the *clean* step, basic data quality tools, such as those described earlier in this chapter, are applied to the data to eliminate or identify the issues with the data. Depending on the *context* or problem domain, the possible uses of the data will inform what additional metadata, such as authentication and other provenance information, is required. For example, an IC application may require a higher level of provenance information. The domain may dictate particular formatting or representation of the data that is best suited. *Categorization* further identifies key properties of interest that can be found in the data. Text analysis is run on the semistructured and unstructured data to identify and extract concepts from the data that are of interest. The particular concepts are dependent on the problem domain. For example, sentiment analysis is used to extract opinions from blog data concerning a product such as a new soft drink. A *correlation* procedure can be applied across the heterogeneous collection of stored data to match and identify data and concepts. This is also called data

---

[*] Derived from discussions in Higgins (2008) and Borne (2010).

integration and entity resolution and can be quite complicated. For example, determining that two references are to the same person (e.g., Mrs. Jones and Ida Jones) is a well-known problem in database management. This step results in various relations being defined, and these are also stored in a repository. Graph databases or triple stores are considered efficient tools for this purpose. Last, in the *catalog* step, the data and its metadata are stored and preserved for its life cycle, and prepared for dissemination (e.g., posting to a shared site, pushing to designated consumers, or indexing for rapid retrieval). Application program interfaces (APIs) are provided for search, extraction, and basic analyses.

The benefits of the data curation process are a reduction in problem-solving time, improved data quality, increased confidence in solutions, reduced time and manual effort to actually perform curation, and the ability to solve problems that were previously too complex or time consuming to solve because of problems with the data.[*] Eventually, digitally curated data can be fed to automated reasoning tools (such as *Wolfram|Alpha* [†] or IBM's *Watson*[‡]) so the analyses can be rapidly accomplished and visualized.

5.15 Big Data Analysis

Many recent advances in analyzing large collections of data have been driven by commercial businesses that are analyzing the online behavior of massive numbers of consumers in an increasing number of applications.[§] For example, Google analyzes users' search and e-mail histories to target advertisements. Facebook is doing similar tracking and analysis. Other companies are looking at Twitter feeds and automatically determining user sentiment. Big Data methods have also arguably transformed some science in a fundamental way through analysis of massive data sets, such as the Sloan Digital Sky Survey, which allows astronomers to search for interesting phenomena in a database of images.[¶] The reader should be cautioned that there

---

[*] Day (2008).
[†] Wolfram|Alpha (2014).
[‡] Fan et al. (2012).
[§] Manyika et al. (2011).
[¶] SDSS (2014).

is much hyperbole surrounding Big Data, with one article declaring "data scientist" to be the "sexiest job of the 21st century,"[*] and some argument over how revolutionary it really is.[†] However, there is no question that many organizations must find ways to deal with high-volume, multiform data that changes rapidly and comes from multiple sources.

There is no universally accepted definition of Big Data, but it has frequently been characterized by means of the "three Vs" of *volume* (large amounts of data; more than can be processed on a single laptop computer or server), *variety* (data that comes in many forms, including documents, images, video, complex records, and traditional databases), and *velocity* (content that is constantly changing, and arriving from multiple sources). To this, a fourth "V" is sometimes added, for *veracity* (data of varying quality).[‡]

The underlying technologies that are enabling the treatment of Big Data include highly distributed and massive computing systems, high-speed communications for both local and global connectivity, and advanced software algorithms that can utilize these computing and communications capabilities. The software algorithms include parallel processing methods, data mining techniques, natural language processing, supervised and unsupervised machine learning algorithms, and new database systems.[§] Much of the software, such as the Hadoop implementation of the Map Reduce Framework for processing large data sets using distributed processing methods, are widely available as open source code.[¶,**] These technologies are rapidly improving and moving toward a high degree of automated content understanding. The 2011 victory of the IBM *Watson* software on the *Jeopardy* television quiz show[††] is an example of the power of these methods to process human language at levels not deemed possible even 10 years ago.

---

[*] Davenport and Patil (2012).
[†] Arbesman (2013).
[‡] Berman (2013).
[§] Orenstein and Vassiliou (2014).
[¶] Dean and Ghemawat (2004).
[**] White (2012).
[††] Best (2014).

Big Data methods are being applied to military applications and are likely to have a large and mixed impact on future operations. On the one hand, they can reduce the manpower and effort necessary to comb through the vast quantities of data and provide meaningful information to the users. On the other hand, they are driving up the demand for ever more data that will further stress the data collection and dissemination systems.

One obvious application for Big Data methods is in the Intelligence, Surveillance, and Reconnaissance (ISR) arena, where the goal is to identify patterns in large amounts of data. For example, methods to analyze video images are being developed that can take over some of the functions of image analysts and only present them with images that contain some interesting behavior.

The U.S. DoD has been investing in Big Data software tools to analyze unstructured data for future warfare applications. According to Kaigham Gabriel, a former research official quoted by *Innovation News*, "The Big Data challenge can be compared to trying to find an object floating in the Atlantic Ocean's nearly 100 billion, billion gallons of water (roughly 350 million cubic kilometers). If each gallon of water represented a byte or character, the Atlantic Ocean would be able to store, just barely, all the data generated by the world in 2010. Looking for a specific message or page in a document would be the equivalent of searching the Atlantic Ocean for a single 55-gallon drum barrel."[*]

Other projects have been looking at methods to convert huge collections of both structured and unstructured data, such as ISR video streams and text documents, into actionable information that can speed up the decision cycle for commanders.[†] Effort has been focused on improving track detection in image analysis, detecting anomalous behavior in video and images such as the placing of Improvised Explosive Devices, and text analysis.

Big Data is rapidly evolving and many of the challenges are being addressed in the research community. In a community white paper from the Computing Research Association (CRA), leading researchers

---

[*] InnoNews (2012).
[†] Schwartz (2011).

have identified a collection of open issues.* The major issues identified include the following:

- Heterogeneity and Incompleteness—How do we deal with missing information?
- Scale—Despite increases in computing power, data volumes are increasing even faster.
- Human Collaboration—Humans can still do some tasks better than machines, and ways to combine the best skills of both are needed.
- System Architecture—We need new designs optimized for moving large quantities of data, and for processing that data, as well as high-level primitives to integrate these systems.

Big Data is still an emerging science, and there is often an over-confidence in the analyses resulting from large collections of social behaviors. For example, the *Google Flu Trends*, an analysis of people's search behavior using terms related to flu, was supposed to predict flu outbreaks faster than the U.S. Centers for Disease Control and Prevention (CDC). However, it was recently pointed out that *Google Flu Trends* overestimated the outbreaks of flu consistently since 2011, showing more than double the amount of flu-related doctor's visits than the CDC.† Others have noticed that false correlations are likely to emerge simply from the large quantity of data being examined. Improved statistical methods that can better deal with large data sets are still required.

The capabilities of Big Data are just beginning to be realized, and new problems that can be tackled with Big Data methods are appearing in many domains. Organizations that can adapt to use these methods are prospering in the current information environment, often beating out those organizations that are failing to change. There are many examples of companies that have adapted to the new information environment and turned it to their advantage. Amazon, FedEx, eBay, and Walmart are but four well-known cases. Translating this success to C2 systems is imperative.

---

* CRA (2012).
† Lazer et al. (2014).

5.16 Onboard Processing

The processing of raw sensor data on the platform carrying the sensor is an important method to reduce the requirements for both communications capacity and the analysis burden of the user, especially for users with limited resources. With the continual reduction in size, weight, and power of computing equipment, there is great potential to perform additional processing. For example, providing moving target tracks rather than complete images directly from the sensor platform would greatly reduce the communication burden and the processing required on the ground.

There is also a continual evolution toward increased autonomy, especially for unmanned air and ground vehicles (UAVs and UGVs). Autonomy requires more powerful onboard processing to accomplish real-time decision making, while still meeting the SWAP limitations. This provides multiple advantages by reducing the demands on human controllers, reducing the load on the communication systems, and reducing the reaction time for the vehicle to maneuver or take action. Recent advances in power sources such as lithium polymer battery technology, solar cells, and fuel cells, as well as advances in microelectronics, such as powerful field programmable gate arrays (FPGAs), are enabling vision processing modules on small UAVs. One example is AeroVironment's *Puma AE*, a hand-launched UAV that has a total weight of 13.5 pounds, carries an electro-optic and infrared camera, has a communication range of 15 km, and can fly for 3.5 hours.* Many new capabilities will require onboard intelligence. Recent research to develop the capability of launching a UAV from the rolling deck of a ship is one example.† It will also be desirable to incorporate even more sophisticated radar, vision, and image processing systems on UAVs.

Another trend is multimission flexibility using architectures that can accommodate various combinations of sensors and processing requirements. This is being stimulated by advances in miniaturized, rugged processing units, advanced real-time operating systems, and improved communications. In addition, the use of open-source software is allowing the incorporation of many more types of UAVs within a single controller system. These traits are also leading to adoption of

---

* AeroVironment (2014).
† Howard (2013).

UAVs for commercial deployments. Spending on UAVs is expected to double over the next decade from about $5 billion in 2013 to $11 billion annually.[*]

One major concern in onboard processing, especially if the raw data is discarded, is that the onboard processing may miss some crucial piece of information, perhaps because the analysis was not as sophisticated as would be possible on the ground. The underlying information would then be lost forever. This is a genuine concern, and it may be desirable to store the raw data for later retrieval if further investigation is desired. However, this introduces a need for additional storage on the system. Given the trends in storage, in particular solid-state storage, this may be less of a problem in the future.

5.17 Human-Computer Interface

Human-computer interfaces have evolved considerably over the last two decades as scientists have increased their understanding of various modes of interaction between persons and machines.[†] The presentation of information to the commander and the ability of the commander to manipulate and filter information can greatly impact effectiveness, and poor interaction designs can hinder effectiveness.[‡] There are many visualization methods that can present the information so as to facilitate rapid understanding of complex situations. Display hardware is improving rapidly in resolution so that smaller displays are both portable and practical. Commanders can use highly capable mobile devices, such as smartphones, that can deliver multimedia information to the tactical edge. Various forms of hands-free operation[§] such as gesture control,[¶] eye tracking,[**] and many applications of speech recognition,[††] allow greater control both in command centers and in the field. Recent technologies such as

---

[*] Howard (2013).
[†] Jacko (2012); Rogers et al. (2011).
[‡] Albers (2011).
[§] Vassiliou et al. (2000).
[¶] Reily and Balestr (2011).
[**] Shumberger et al. (2005).
[††] Pigeon et al. (2005).

Google Glass can serve as platforms for augmented reality* applications, so that soldiers can have wearable displays† that provide multimedia information to their headsets while on the move. Software practices and frameworks have improved so that rapid development of interfaces can be accomplished using standard methods such as widgets and deployment methods such as App stores. Recent efforts of the Defense Information Systems Agency (DISA) to improve the Global Command and Control System–Joint (GCCS-J) have included open-source frameworks, such as *Ozone Widget Framework*, to make the system more configurable and consistent.‡ All of these advances are increasing the effectiveness of C2 systems but are also stoking the need for more data.

In some respects, as we discussed in Chapter 4, our society has become accustomed to operating in a connected, always-on, information-rich environment, at homes, schools, and workplaces. Today's generation of soldiers, as we also discussed in Chapter 4, are products of this society, and many are familiar and comfortable with the multitasking, highly stimulated environment enabled by ever-present smartphone and Internet access. However, as mentioned earlier, there are various studies beginning to look at the cognitive effects of trying to deal with Ubiquitous Data. The DoD and many others are recognizing the negative effects of data overload, and research is beginning to focus on the problem of how to cope with it. Some relevant results from cognition research include the following§:

- Brain wave activity from simulation of military drone operators at the Nevada Air Force Base shows spikes in theta waves as load increases that indicate overload.¶
- Recent research has suggested that a human analyst eventually reaches a physical threshold for how much information he or she can process at one time, ultimately determined by blood flow to the brain.**

---

* Azuma (1997); Avery et al. (2010).
† Consider, for example, Behringer et al. (2000a,b).
‡ DISA (2012).
§ Shanker and Richtel (2011).
¶ Hsu (2011); Parasuraman (2011).
** Parasuraman and Wilson (2008); Parasuraman (2011).

- People who have grown up constantly switching attention ("multitasking") may have trouble focusing. In experiments, soldiers operating tanks can miss seeing nearby targets when looking at remote video feeds.[*] Heavy multitaskers do significantly worse on filtering out irrelevant information. In addition, multitaskers take longer to switch between tasks.[†,‡]
- Frequent interruptions such as from episodic incoming e-mail messages increase stress.[§]
- Research has shown that sufficient downtime is necessary for the brain to properly process inputs, and that processing too much data can leave people fatigued.[¶] Another study has shown that people learn better after taking a walk in nature (lesser stimulation) versus a walk in a city.[**]

In the opinion of experts at the National Institute of Drug Abuse, one can "compare the lure of digital stimulation less to that of drugs and alcohol than to food and sex, which are essential but counterproductive in excess" (Nora Volkow, director of the National Institute of Drug Abuse).[††] Modern C2 organizations will need to recognize the benefits and limitations of always-on connectivity and Ubiquitous Data.

5.18 Technology and the C2 Organization

The technologies discussed above, and many others, will continue to automate the tasks required to process and effectively use the massive streams of data that are now available to the C2 organizations. How will these changes impact the C2 organization? The newer technologies such as digital data curation, onboard processing, and Big Data should reduce the response time to process large amounts of data in

---

[*] Hairston et al. (2012).
[†] Richtel (2010a).
[‡] Ophir et al. (2009).
[§] Richtel (2010a).
[¶] Richtel (2010b).
[**] Berman et al. (2008).
[††] Richtel (2010a).

the OODA loop* by reducing the manual components, and thereby supporting better situation awareness. The ability to disseminate and share the information can be increased with information interoperability enhancements. Visualization and cognitive studies will better inform the best use of personnel and work conditions and human-computer interfaces. These advances all support distributed, decentralized, net-enabled C2 Approaches empowering many layers of command, with widespread information dissemination and broad interaction patterns. We discuss those concepts further in the next chapter.

5.19 Concluding Remarks

During a conference on C2 held at the Institute for Defense Analyses in 2011,† an officer from the U.S. Marine Corps recounted an illustrative anecdote. He had participated in the 2003 march to Baghdad. For about three weeks during that time, he was disconnected from his e-mail. On arriving in Baghdad, he found over 1600 messages waiting for him. He remembers being angered, for two reasons: first, that he could have used some of those messages; second, that even if he had received all his messages in a timely fashion, he might not have been able to separate the useful from the useless, particularly under the stress of a combat advance. It is not only the proliferation of video sensors and UAVs that create a data deluge. A data deluge can easily happen even with low-bandwidth text messages.

The Marine officer's dilemma has become commonplace in both the military and civilian worlds. Megatrend 3, Ubiquitous Data, is a direct result of Megatrend 2 (the Robustly Networked Environment). As a result of Megatrend 2, anyone can create and disseminate information on a scale that heretofore would not have been possible, let alone economically viable. The cheaper and more widespread the technology, the worse the problem potentially becomes. Advanced and cheap technology, as discussed above and in Chapter 4, empowers adversaries to create more information and disinformation that

---

* Observe, Orient, Decide, Act (OODA) is a paradigm developed by John Boyd, extensively used in the C2 community. See Boyd (1995).

† IDA/OASD (R&E) Conference on Commercial Technologies in C2, held at the Institute for Defense Analyses, Alexandria, VA, on May 10, 2011. Proceedings are in an IDA internal document.

must be sorted and analyzed. New commercial technologies eventually make their way into the military and its coalition partners in some form, and serve to increase both the coalition's information exchanges and each participant's internal flows and transactions, with the attendant potential for information overload.

We see in Chapters 6 and 8 that Megatrend 3, Ubiquitous Data, is also a result of Megatrend 4, the availability of alternative, net-enabled organizations—which are enabled in turn by Megatrend 2, the Robustly Networked Environment, and often demanded by Megatrend 1, Big Problems. In some situations, every soldier is a potential source of information in the form of text, image, and video that could be critical to the mission, and that must be processed, viewed, and digested.

Much of the discussion in this chapter has focused on data and information quality. Data and information quality are central to understanding some of the implications of Ubiquitous Data for C2. Chapter 8 will examine the empirical evidence for the effects of data and information quality, and the relationships between them and the organizational designs and C2 Approaches that we will discuss in the next chapter.

6

MEGATREND 4: NEW FORMS OF ORGANIZATION

6.1 Introduction

Megatrend 1 (Big Problems) may involve, as we have seen, complex endeavors coupled with complex enterprises. Complex, dynamic, uncertain environments cannot always be sufficiently understood and controlled using rigid, Industrial Age hierarchical approaches. Big Problems can force alternative organizational forms to emerge, and those alternatives are often more decentralized and net enabled. As we see below, such alternative organizations and approaches are not necessarily new ideas; the business literature has discussed them for decades, and military concepts of mission command date back to the 19th century. However, Megatrend 2 (Robustly Networked Environment) makes many such approaches increasingly feasible, in an ever-growing number of situations.

Megatrend 2 also enables a hitherto unanticipated set of new adversaries and competitors to adopt decentralized approaches, organizations, and operations with considerable sophistication. To the extent that these approaches make adversaries nimbler and harder for legacy establishments to comprehend, an effective counter may often require increased agility on the part of more established entities. Megatrend 2 is not simply an enabler of Megatrend 4; there can be a direct cause-and-effect relationship between the two.

6.2 C2 Approaches

6.2.1 Notional Organizational Examples

Burgess and Fisher (2008) have developed a useful framework for thinking about net-centric, decentralized Command and Control

LEVEL NUMBER	KEY FUNCTION	CONVENTIONAL DESCRIPTOR—FINE	CONVENTIONAL DESCRIPTOR—COARSE
1	What is the problem? Who are we? Who are our enemies? Who are our allies? Etc.	National Strategic	Strategic
2	What can we do about it? Who plays? Who pays?	Military Strategic	
3	How and when will we deal with it? When? Where? Using what resources?	Operational	Operational
4	Who will execute? Team formation, preparedness, etc.	Operational/Tactical	
5	How will we execute? Targets for effects	Tactical	Tactical
6	Actions required (Down to individual level)	Individual	

Figure 6.1 Burgess and Fisher's Command Level Framework, augmented with other conventional descriptors.

(C2). They have delineated a number of generic command functions of varying levels of abstraction (Figure 6.1). These can map onto corresponding hierarchical structures but need not necessarily do so. At higher levels of abstraction, considerations include the broadest nature of the problem at hand, the identity of self, the identity of enemies, and so on. At the lowest levels of abstraction are determination and execution of immediate actions by individuals or very small groups. These levels of abstraction have been given various conventional names in the general C2 literature, such as "strategic" at the highest levels of abstraction, "operational" at the middle levels, and "tactical" at the lowest. There can be some debate about when exactly "operational" shades into "tactical," and sometimes finer gradations of terminology are used. Some of the common variations are captured in Figure 6.1.

In a traditional hierarchical, centralized C2 structure, the higher-level command functions are reserved for higher levels in the hierarchy, as shown in Figure 6.2a. When the level of abstraction of the command function does not correspond directly to a hierarchical level, various results are possible. We show only a few examples here for the sake of illustration. Figure 6.2b shows an example of the proverbial "6000-mile-long screwdriver," wherein an individual highly placed in the hierarchy is performing lower-level command functions and engaging with people much farther down the chain. This type of micromanagement is not normally considered a desirable *modus operandi*, as it removes initiative from those closest to the situation at hand. Figure 6.2c shows an example of Charles Krulak's "strategic

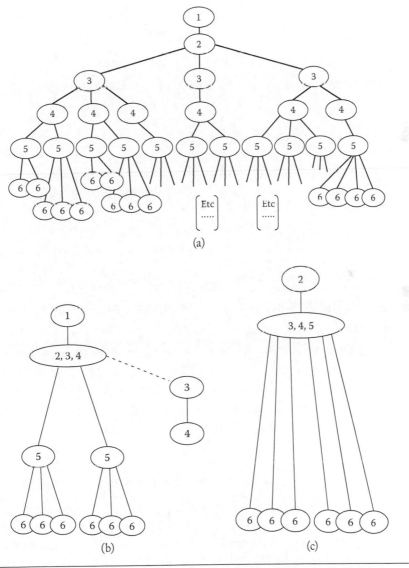

(a)

(b)   (c)

Figure 6.2   Notional examples of (a) hierarchical C2, (b) micromanagement, (c) a "strategic corporal" or small autonomous unit, and (d) a modern networked force. (Adapted from Burgess and Fisher, 2008. With permission.) (*continued*)

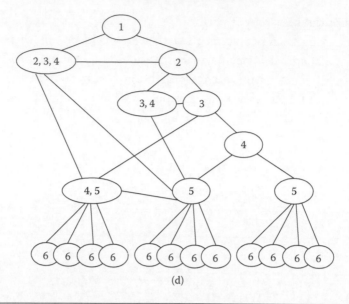

(d)

Figure 6.2 (*continued*) Notional examples of (a) hierarchical C2, (b) micromanagement, (c) a "strategic corporal" or small autonomous unit, and (d) a modern networked force. (Adapted from Burgess and Fisher, 2008. With permission.)

corporal in a three-block war."* The proverbial "corporal" performs many high-level command functions and autonomously directs his small unit, whose members might also be operating quasi-independently. Figure 6.2d shows a more complex web of command relationships, as might be conceived for a modern networked force. Note that the diagrams in Figure 6.2 depict command relationships, and not just information flows. If Figure 6.2b, for example, were depicting information flows and not command relationships, it might not be indicative of micromanagement but of a complex pattern of information sharing that may be desirable.†

---

* Krulak (1999).

† Consider, for example, the U.S. military's Strategic Knowledge Integration Web (SKIWeb), a Web-based asynchronous collaboration system introduced at the U.S. Strategic Command (STRATCOM) under the leadership of General James Cartwright. At the end of 2009, the system had 28,000 users throughout the Department of Defense, at all levels of command. Any user, from a general to a front-line soldier, could post information, and that information was accessible to all. Users could also expand, correct, and comment on others' posts. The system stimulated a nonhierarchical flow of information. See Cartwright (2006); Wyld (2007); Soknich (2009); Vassiliou (2010).

6.3 C2 Approach Space

Alberts and Hayes (2006) describe a succinct and effective categorization of approaches to C2. In their C2 "Approach Space," shown in Figure 6.3, the dimensions are as follows:

- *Information distribution* (tightly controlled or broadly disseminated)
- *Patterns of interaction* between actors (tightly constrained or not)
- *Allocation of decision rights* (unitary or peer to peer)

In plain English, the Approach Space considers who knows what, who can talk to whom, and who can decide to act.

Classic, centralized, hierarchical C2 systems tend to have unitary allocation of decision rights, constrained patterns of interaction, and tightly controlled information flows. Thus, the organization depicted in Figure 6.2a would plot in the lower left of the cube. The upper-right vertex of the cube represents a theoretical "edge organization," with the broadest possible information distribution, patterns of interaction, and allocation of decision rights. Thus, moving along the diagonal of

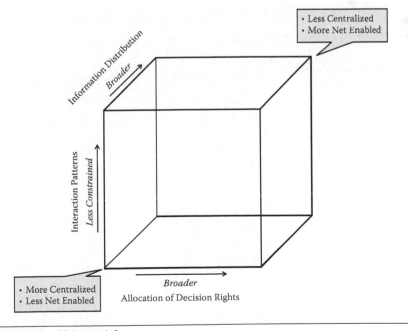

Figure 6.3 C2 Approach Space.

the cube, from lower left to upper right, indicates varying degrees of balanced decentralization.

Real organizations will occupy regions in the cube, which represents a conceptual continuum. As we plot such regions throughout this book, it is important to note that the depiction is qualitative and notional rather than strictly quantitative. The plots are meant to illustrate concepts and illuminate relationships in a broad sense. Figure 6.4 shows the notional positions of the organizations in Figure 6.2.

Agile C2 systems will occupy whatever part of the cube is most appropriate for their mission.

6.4 C2 Approach Space for Collective Endeavors

With some modification, the Approach Space can be extended to apply to collective endeavors, which involve groups of organizations, each of which may have its own culture and C2 Approach. Complex endeavors such as major disaster relief often involve such

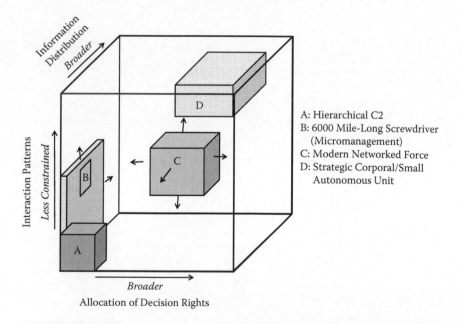

A: Hierarchical C2
B: 6000 Mile-Long Screwdriver
   (Micromanagement)
C: Modern Networked Force
D: Strategic Corporal/Small
   Autonomous Unit

Figure 6.4 Representations of (A) hierarchical C2, (B) micromanagement, (C) a modern networked force, and (D) a "strategic corporal" or small autonomous unit in the C2 Approach Space. Positions are illustrative and not intended to be taken as precise.

collectives, as do military coalitions. The dimensions of the cube now become

- *Information distribution among entities* in the collective (tightly controlled or broadly disseminated)
- *Patterns of interaction among entities* in the collective (tightly constrained or not)
- *Allocation of decision rights among entities* in the collective (unitary or peer to peer)

Each entity in the collective may have its own approach (its own "cube") internally; the collective Approach Space describes how the various entities interact with *one another*.

Figure 6.5 illustrates the Approach Space for collective endeavors, along with names for characteristic types of approaches to collaboration among the entities, which have been characterized as follows by Alberts et al. (2010):

- *Conflicted C2*: There is no collective objective. The only C2 that exists is that exercised by the individual contributors over their own forces or organizations.

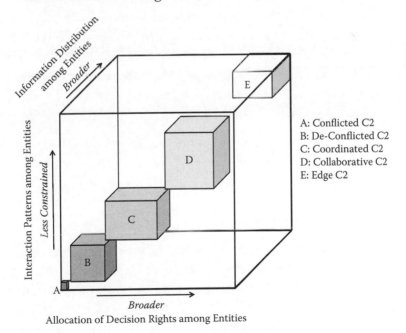

A: Conflicted C2
B: De-Conflicted C2
C: Coordinated C2
D: Collaborative C2
E: Edge C2

Figure 6.5 C2 Approach Space for collective endeavors. Positions are illustrative and not intended to be taken as precise.

- *De-Conflicted C2*: Entities interact just enough, share just enough information, and give up just enough decision rights to others so as to avoid adverse cross-impacts between and among the participants in the collective.
- *Coordinated C2*—Entities now do more than simply modify their intent, plans, and actions to avoid potential conflicts. In Coordinated C2 they develop some degree of common intent and an agreement to link actions in the various plans being developed by all the entities. Entities (1) seek mutual support for intent, (2) develop relationships and linkages between and among entity plans and actions to reinforce or enhance effects, (3) pool nonorganic resources to some extent, and (4) share more information to improve the overall quality of information.
- *Collaborative C2*—Entities move beyond shared intent and now seek to collaboratively develop a single shared plan. They do this by (1) negotiating and establishing collective intent and a shared plan, (2) establishing or reconfiguring roles, (3) coupling actions, (4) sharing nonorganic resources, (5) pooling nonorganic resources to some extent, and (6) increasing social interactions to augment shared awareness.
- *Edge C2*—Entities achieve self-synchronization. The collective becomes a robustly networked collection of entities with widespread and easy access to information, extensive sharing of information, rich and continuous interactions, and the broadest possible distribution of decision rights.

Chapter 8 discusses many experimental instantiations of these collective approaches and their implications.

6.5 Business Organizations

There are some commonalities between the theory of C2 Approaches and the organization theory that has been explored for decades in the business literature. Although there are certainly differences between entities organized primarily for profit and those accomplishing military missions or disaster relief, it is worthwhile to explore how some concepts from the business literature fit into the C2 Approach Space.

6.5.1 Mintzberg's Organizational Archetypes

Mintzberg (1979, 1980) identifies several organizational archetypes that arguably span most of the space of observed business entities. Described in simple terms, the most important of Mintzberg's types for our purposes are as follows[*]:

- *The Simple Structure*—This is an organization with vertical lines of authority, but it is also relatively flat, without multiple layers of management. Mintzberg identifies direct supervision as the key coordinating mechanism. Authority is concentrated at the top, but there is also considerable flexibility. Internal organization forms organically according to functional needs, and there are few formal supporting structures. Small companies and some startups often fall into this category.

- *The Machine Bureaucracy*—This is a hierarchical organization with a high level of standardization of tasks and procedures. Mintzberg identifies "standardization of work" as the key coordinating mechanism for this type of organization. Traditional industrial companies engaging in mass production have often been organized this way, as have service companies with work tending to simple and repetitive tasks. This form is also common among government agencies, including the business operations of military establishments. Machine bureaucracies are best suited to relatively simple and stable external environments. Despite the fact that military establishments must sometimes deal with environments that are anything but stable or simple, classic hierarchical C2 has much in common with the machine bureaucracy.

---

[*] Mintzberg's other archetypes, introduced in various publications, include (1) the Divisional Form, (2) the Missionary Organization, and (3) the Political Organization. The Divisional Form is similar to a collective of several entities, for example, a collection of organizations each of which is a Machine Bureaucracy. Thus, we do not consider it a primitive. The Missionary Organization is one in which ideology is key and the main coordinating mechanism is standardization of norms. From the point of view of the C2 Approach Space, it can probably be subsumed by a number of the other archetypes. The Political Organization, as its name implies, is dominated by politics. It has no prime coordinating mechanism and can assume any number of forms.

- *The Professional Bureaucracy*—This is an organization typically focused on highly trained professionals who can work with a degree of autonomy. Its bureaucratic standardization is focused on human resources aspects rather than on formalizing the execution of tasks; Mintzberg identifies "standardization of skills" as the key coordinating mechanism. For example, the organization may expend considerable effort on the definition of required positions and capabilities. Professional bureaucracies may evolve to confront environments that are stable but complex. Mintzberg gives accounting firms and craft manufacturing firms among his examples of the types of companies that develop professional bureaucracies.

- *The Adhocracy*—An adhocracy is a flexible organization without a large amount of formal structure. Typically, employees of adhocracies will be highly skilled and able to work autonomously. Teams may form and organize flexibly according to the requirements of the project at hand. The vertical and horizontal flow of authority may also vary depending on the project, with reconfigurable forms of matrix management. Mintzberg identifies "mutual adjustment" as the key coordinating mechanism. Adhocracies are suited to complex and rapidly changing environments, in which a more formal organization may be too slow to respond. Some adhocracies may do similar work as professional bureaucracies. Adhocracies may be observed in a wide variety of companies and industries, from aerospace companies to professional services firms and startup companies. The adhocracy is the archetype closest to the net-enabled C2 "edge organization."

Mintzberg discussed several variables ("design factors") to characterize the organizational archetypes. The subset below represents the important parameters that can be mapped to the C2 Approach Space*:

---

* The others are as follows: (1) Training and Indoctrination, by which skills and knowledge are standardized and transmitted; (2) Unit Grouping, the bases by which job positions are clustered into units and various levels of super units; and (3) Unit Size.

- *Centralization* is concerned with the concentration of authority and the allocation of decision rights.
- *Specialization* is the parameter describing the division of labor. Horizontal specialization concerns the number of tasks and their breadth, while vertical specialization refers to the performer's level of control over the tasks. Generally, "unskilled" jobs have high degrees of both vertical and horizontal specialization, while skilled ones have high horizontal specialization and less vertical.
- *Formalization* defines the extent of formal specification and standardization of work processes. Unskilled jobs tend to be more formalized.
- *Liaison Devices* are the means of horizontal interaction and mutual adjustment. These might range from very informal exchanges to highly formalized matrix structures.
- *Planning and Control Systems* refers to how outputs are standardized and managed. This may consist of action planning (predetermination of outputs) or performance control (forensic measurement of effects and outputs).

Alberts and Nissen (2009) analyze Mintzberg's parameters and find that while there is no direct correspondence to the axes of the C2 Approach Space, combinations of them can be mapped onto each of the space's three axes. Perhaps the most obvious correspondence is between Centralization and the Allocation of Decision Rights. However, Horizontal Specialization and Vertical Specialization, which directly impact how people can accomplish their tasks, also affect the Allocation of Decision Rights. One could also argue (Alberts and Nissen do not) that Formalization has some effect as well. Liaison Devices, along with Planning and Control Systems, clearly affect both Patterns of Interaction and Distribution of Information. Patterns of Interaction are also affected by Formalization.

Using these correspondences, we may construct a rough representation of Mintzberg's archetypes on the C2 Approach Space (see Figure 6.6).

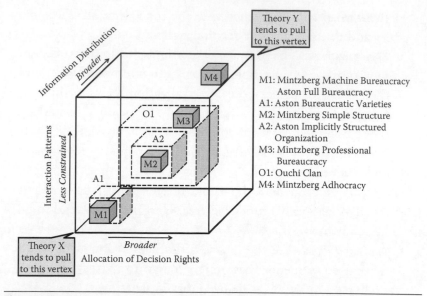

Figure 6.6   Business organizational models from organization and management theory in the C2 Approach Space. Positions are illustrative and not intended to be taken as precise.

6.5.2 *The Aston Studies*

The "Aston Studies" were a seminal set of investigations of the organizational structure of a large number of entities in England, undertaken in the 1960s. They considered 64 different parameters that the management literature up to that time had identified as important in defining the characteristics of organizations. Through mathematical analysis, the studies were able to reduce the plethora of parameters to three underlying dimensions as described by Pugh et al. (1969):

- *Concentration of Authority*—This is largely self-explanatory, and related to Mintzberg's *Centralization*.
- *Structuring of Activities*—This is the degree of formal definition of employee roles. This dimension combines aspects later formulated by Mintzberg in his *Formalization* and *Specialization* parameters.
- *Line Control of Workflow*—This refers to the degree of control that line workers can exercise over their activity, as opposed to having those activities prescribed by "impersonal" procedures. It is a measure of worker autonomy and similar to Mintzberg's later *Formalization* parameter.

As with Mintzberg's parameters, the Aston dimensions do not map directly onto the C2 Approach Space. The clearest link is between *Concentration of Authority* and the C2 Approach Space's *Allocation of Decision Rights*. However, *Allocation of Decision Rights* is influenced by both the other Aston dimensions as well. *Line Control of Workflow* and *Structuring of Activities* are both linked to *Patterns of Interaction*. The least clear link is between the Aston dimensions and the C2 Approach Space's *Distribution of Information*. Arguably, there is some relationship between the *Distribution of Information* and the *Structuring of Activities*.

The Aston Studies identified a number of organizational archetypes before Mintzberg enunciated his. These tend to span a narrower region of organizational space. Three of the Aston archetypes are bureaucracies.* The *Workflow Bureaucracy* has a high level of task standardization and is described as applying typically to large-scale manufacturing or "big business." The *Personnel Bureaucracy* has strict standardization of human resource activities such as recruiting, selection, and dismissal, but less strict standardization of workflow. Examples in the Aston Studies are a municipal education department, a regional division of a government ministry, or a smaller branch factory of a large corporation. The *Full Bureaucracy* combines aspects of the Workflow and Personnel bureaucracies, and is thus the most bureaucratic of all. The Aston Studies give a manufacturing branch factory of a central government department as an example. All three of these bureaucratic archetypes are probably similar to Mintzberg's *Machine Bureaucracy*, or some definition of it with somewhat enlarged scope. The *Personnel Bureaucracy* shares some characteristics with Mintzberg's *Professional Bureaucracy* but is probably not as decentralized overall if mapped to the C2 Approach Space.

Another Aston archetype is the *Implicitly Structured Organization*. This is an organization without large amounts of formal structure and with a relative decentralization of decision rights. The Aston studies assigned this description to relatively small factories in their sample, where the original owners also still had considerable influence.

---

* The Aston Studies (Pugh et al., 1969) also identify some archetypes as being notional way stations on their way to becoming one of the main ones. For example, they define a "Pre-Workflow Bureaucracy," a "Nascent Workflow Bureaucracy," and a "Nascent Full Bureaucracy"; Walker et al. (2009).

This archetype overlaps somewhat with Mintzberg's *Simple Structure*. Notional positions of the Aston archetypes in the C2 Approach Space, in comparison to Mintzberg's archetypes, are shown in Figure 6.6.

6.5.3 Others

The literature on organization and management theory is vast, and we make no attempt to cover all of it, particularly since the Mintzberg archetypes provide an adequate delineation of the overall organizational space. However, we can mention the organizational categories proposed by William Ouchi[*]: the *Market-Based Organization*, the *Bureaucracy (Hierarchy)*, and the *Clan*. Of these, the most interesting is the *Clan*, where social mechanisms produce a strong sense of community among employees, allowing effective functioning with some measure of devolved authority and without a large amount of formal structure. There are some similarities between Ouchi's *Clan* and Mintzberg's *Professional Bureaucracy*, *Simple Structure*, and perhaps even *Adhocracy*. There are also possible similarities with the Aston *Implicitly Structured Organization*. Ouchi's *Market* organization has some commonality with the Mintzberg *Professional Bureaucracy*.

6.5.4 Theory X, Theory Y, and Trust

It is also worth discussing Douglas McGregor's "Theory X" and "Theory Y." In his 1960 book, *The Human Side of Enterprise*, McGregor identified assumptions he believed were central to traditional management (Theory X) and proposed a different set of assumptions (Theory Y) that he considered more enlightened and more potentially effective.

Theory X, as enunciated by McGregor, considers humans fundamentally lazy and assumes they will avoid work if they can. It also assumes that humans do not desire responsibility and will avoid it. Furthermore, they will always put their own needs above those of the organization. Thus, the job of management is to direct people, motivate them, coerce them, and control them, modifying their behavior to

---

[*] Ouchi and Price (1978).

meet the goals of the organization. McGregor believed that this style of management bred mutual distrust, antagonism, and inefficiency.*

To counter this theory of management, McGregor proposed a different set of assumptions, embodied in his "Theory Y." He posited that individuals actively seek to work, and work can be a source of satisfaction and self-fulfillment. People do not necessarily shun responsibility and may seek it under the right conditions. They are also more generally able to innovate than Theory X would admit. Thus, close supervision and coercion are not always the best way to get productive effort out of employees. A key role of management, then, is to help people develop their natural capacity for assuming responsibility and self-direction. McGregor did not necessarily believe that a "pure" Theory Y organization could be created, but he argued that Theory Y assumptions could increase the effectiveness of management.

McGregor outlined approaches to management that he considered consistent with Theory Y. These included delegation, participative management, expansion of the scope of most jobs, and decentralization of authority. Thus, in theory at least, McGregor argued for moving from the "hierarchical" vertex of the C2 Approach Space toward the "Edge" vertex. As we see in our discussion of mission command in Section 6.7, having trust in the intentions, motivations, and training of one's workforce, whether they are conventional employees or military personnel, is crucial to being able to broaden the allocation of decision rights, patterns of interaction, and distribution of information.

6.5.5 Are Hierarchies and Bureaucracies Things of the Past?

The short answer is, "no," despite the rather large amount of published business literature decrying the traditional hierarchy and all its flaws.†

---

* Assumptions similar to those behind Theory X are often identified with the early 20th century management theorist Frederick Winslow Taylor, father of "Scientific Management" (Taylor, 1911). However, this is an unjust oversimplification of Taylor's thinking (Brogan, 2011; Blake and Moseley, 2011; Bobic and Davis, 2003). While Taylor did not explicitly enunciate propositions identifiable as "Theory Y," neither did he consider workers as mere lazy and unmotivated cogs in a machine.

† Consider, for example, Hamel (2011); HBR (2010); Kotter (2012); Seddon (2005). These and other publications call for new and more agile organizational forms in business, but we do not mean to imply that they all claim hierarchies must be eliminated.

Realistically, we can expect that hierarchies will always exist. In business, untold numbers of products have been developed and brought to the consumer, and untold fortunes have been made by hierarchies. Hierarchies are good at enforcing accountability,* and legal doctrines in society will almost always mandate some form of hierarchical organization for both civilian and military establishments. Hierarchies are also well adapted to some problems, particularly ones that are decomposable into parts that can be solved in isolation.

However, hierarchies need not be absolute, and they need not represent the only alternative at all times, even in a single organization. They need not occupy the extreme lower-left corner of the Approach Space. They can often coexist with more flexible structures, with broader allocation of decision rights, broader patterns of interaction, and broader information distribution. In the information age, even a classic hierarchy is different from the classic hierarchy of the past: it is likely better networked and has broader information distribution, even despite itself.

The military establishments that have employed mission command, as we discuss in Section 6.7, have not eliminated their hierarchies or their chains of command. Military and other organizations can create teams and even large enterprises that adopt more flexible approaches when the situation warrants it.

6.6 Decentralized Organizations

In their book *The Starfish and the Spider*,† which argues for the effectiveness of decentralized networked organizations, Ori Brafman and Rod A. Beckstrom offer the Apache Indians of Mexico and the Southwestern United States as an example. The Apaches managed to repel the more technologically advanced Spanish Army for two centuries, and often confound the U.S. military as well.‡ The various Apache groups had a relatively egalitarian society and a diffuse political organization. Leaders of Apache bands were respected men

---

* Jaques (1990).
† Brafman and Beckstrom (2006).
‡ The leader known as Victorio fought the U.S. and Mexican militaries, which vastly outnumbered his forces, for nearly a year before being killed (Watt, 2011).

of superior ability, but their positions were not permanent, their word was not absolute, and they had little coercive power over their followers. The Apaches followed their leaders voluntarily.[*] There was no central location for an adversary to attack and destroy. As Brafman and Beckstrom put it, "Apache decisions were made all over the place. A raid on a Spanish settlement, for example, could be conceived in one place, organized in another, and carried out in yet another. You never knew where the Apaches would be coming from. In one sense, there was no place where important decisions were made, and in another sense, decisions were made by everybody everywhere."[†] Under the pressure of Spanish attacks, the Apaches scattered and became even more flexible and difficult to defeat. The Apaches thus employed edge-like behavior at times and their own version of agile C2.

Some potential adversaries of the west in the 21st century are arguably already behaving in a similar fashion. This is particularly true of some terrorist networks. Consider Al Qaeda. Some observers have commented that before the attacks of September 11, 2001, Al Qaeda was a hierarchical and even bureaucratic organization,[‡] displaying agility and transforming into a much looser and decentralized network in answer to the pressure exerted by the U.S. response to the attacks.[§] However, to call even pre–September 11, 2001 (9/11) Al Qaeda a hierarchy of the conventional type is probably an exaggeration. Even before 9/11, Al Qaeda displayed much of the character of a relatively decentralized network.[¶]

Al Qaeda appears to have executed the attacks of 9/11 in a relatively decentralized fashion.[**] Osama Bin Laden[††] is thought to have known of the plan and blessed the operational concept at a high level, but he is not believed to have known the operational details, such as flight numbers. Importantly, his subordinates understood his intent. They were trained and empowered to carry out the operation and fulfill that intent on their own. Al Qaeda operatives have shown a

---

[*] Goodwin (1935); Basehart (1970); Kaut (1974); Tyler (1965).
[†] Brafman and Beckstrom (2006), pp. 20–21.
[‡] Hoffman (2006); Michael (2012).
[§] Ibid.; McGrath (2011).
[¶] McGrath (2011).
[**] Zwikael (2007); Saunders (2002).
[††] Osama bin Mohammed bin Awad bin Laden (1957–2011).

capability to self-synchronize, through unity of effort (a shared fundamentalist faith), a clear understanding of the commander's intent, and common rules of engagement. They have also made effective use of available information and communication technologies, employing the Internet and cell phones to develop shared awareness of intelligence and, ultimately, knowledge superiority. On September 11, 2001, they knew their adversary rather better than their adversary knew them.

Some other terrorist organizations also show edge-like behavior. One example is the grassroots jihadi network responsible for the Madrid bombings of 2004.* Despite some weaknesses, the group managed to function in an autonomous and agile manner, without a continual need to consult senior levels. The network was an ad hoc grouping with a complex leadership web, driven by a shared intent to carry out a terrorist bombing. Other examples of edge-like organizations include right-wing extremist groups in Germany and the United States. In fact it was an American right-wing extremist, Louis Beams, who enunciated, in 1992, the concept of "leaderless resistance."† There is evidence that right-wing groups followed this as a guideline at least to some extent.‡ Such groups, while not as apocalyptically successful as Al Qaeda, have made effective use of the Internet as a C2 medium.§

Not all terrorist organizations or insurgencies behave as net-centric edge organizations. The "traditional" terrorist groups of the 20th century, such as the Provisional Irish Republican Army (IRA) and the Basque separatist organization *Euskadi Ta Askatasuna* (ETA), had tighter hierarchical C2 architectures. The IRA and its splinter groups moved toward more edge-like behavior under pressure from law enforcement.¶ This is shown in Figure 6.7.

It is sometimes stated that, "It takes a network to defeat a network."** This seems like a reasonable assertion although difficult to prove

---

* Jordan et al. (2008).
† Jones (2007); Shields (2012).
‡ Shields (2012).
§ Concepts similar to leaderless resistance have in fact also been explicitly enunciated by some members of Al Qaeda. Michael (2012) gives the example of the captured Syrian operative Abu Musab al-Suri, who proposed a "jihad of individual terrorism."
¶ Jackson (2006).
** Consider, for example, Smith (2006).

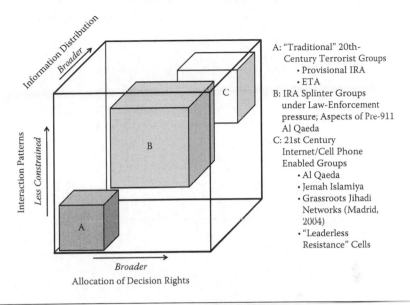

A: "Traditional" 20th-
   Century Terrorist Groups
   • Provisional IRA
   • ETA
B: IRA Splinter Groups
   under Law-Enforcement
   pressure; Aspects of Pre-911
   Al Qaeda
C: 21st Century
   Internet/Cell Phone
   Enabled Groups
   • Al Qaeda
   • Jemah Islamiya
   • Grassroots Jihadi
     Networks (Madrid,
     2004)
   • "Leaderless
     Resistance" Cells

Figure 6.7   Estimated notional positions of some adversary entities in the C2 Approach Space. Positions are illustrative and not intended to be taken as precise.

definitively. If an adversary organization's net-centric behavior makes it agile and flexible, it seems likely that one will be better positioned to defeat it if one also adopts those attributes. An instructive example may be found in the experience of the Israeli Defense Forces (IDF) at Nablus in 2002.* The IDF faced a loose confederation of organizations including Hamas (see Figure 6.8), Palestinian Islamic Jihad, some security forces from the Palestinian Authority, and street gangs. The groups coordinated with each other to a limited extent, but during the fighting they were autonomous and self-synchronizing. In order to fight them, the IDF formed small networks of its own, giving field commanders considerable autonomy. The small units exchanged information efficiently, both horizontally and vertically. The IDF engaged quickly and then withdrew, in a largely successful operation. The strategy required higher-level commanders to accept the autonomy of lower-level commanders, and required commanders in general to accept more questioning from subordinates.

The IDF's experience in the 2006 war with Hezbollah was not as positive. During this conflict, Hezbollah acted in a complex fashion,

---

* Jones (2007).

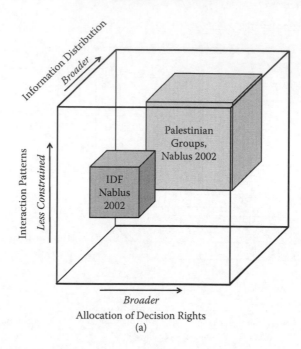

(a)

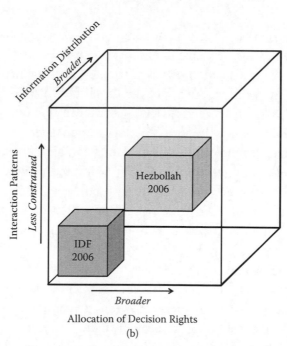

(b)

Figure 6.8   Estimated notional positions in the C2 Approach Space of (a) the Israeli Defense Forces and Palestinian groups at Nablus in 2002, and (b) the Israeli Defense Forces and Hezbollah in 2006. Positions are illustrative and not intended to be taken as precise.

blending conventional and irregular warfare in a manner that has been labeled "hybrid" (see Figure 6.8).[*] The IDF's strategy involved conventional air operations, followed by a somewhat belated ground response when those did not fully succeed. Hezbollah's C2 had both hierarchical and distributed elements. There was apparently a formal chain of command, operating from command posts with fairly sophisticated equipment, including landline cables and encrypted radios.[†] However, Hezbollah also employed a distributed network of small units acting with considerable autonomy, displaying unity of effort and a degree of self-synchronization.[‡] Hezbollah also used a somewhat decentralized media strategy, without the long message-approval cycles present in conventional militaries. Members understood the intended message and many were equipped with relatively inexpensive new media technologies enabling them to get it out fast, and significantly outflank Israel in the propaganda war.[§]

6.7 Mission Command

6.7.1 Introduction

Many western military establishments, including that of the United States, have adopted some form of mission command[¶] as part of their stated doctrine. Mission command is a C2 framework of varying degrees of decentralization, potentially investing lower levels of the hierarchy—which remains essentially intact—with considerable autonomy. It is not synonymous with net-enabled decentralized C2 and does not in itself guarantee the development of agile organizations. However, mission command can provide a fertile soil from which agile C2 Approaches may grow; and net-centric technologies can in turn be used to facilitate mission command.

---

[*] Jordan (2008).
[†] Biddle and Friedman (2008).
[‡] Cordesman (2006); Rourke (2009).
[§] Collings and Rohozinski (2009).
[¶] Wittmann (2012); Shamir (2011).

6.7.2 Essential Features of Mission Command

In mission command, commanders are expected to issue only the most essential orders, providing objectives and general instructions, and establishing command intent. Subordinates have considerable autonomy to develop tactical details suited to the particular real situations they face. While detailed, strictly hierarchical C2 seeks, perhaps in vain, to reduce battlefield uncertainty, mission command embraces the uncertainty and empowers subordinates to deal with it. As shown in Figure 6.9, mission command involves broadening the allocation of decision rights to varying degrees and to varying extents within an organizational structure. As mission command moves from theory to practice, there is often a concomitant broadening of interaction patterns and information distribution.

Mission command requires a large amount of trust between commanders and subordinates at all levels. It can also require a considerable amount of situational awareness and coordination to avoid complete disintegration of the effort.

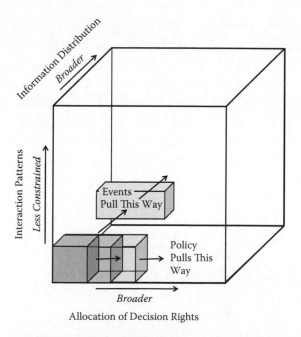

Figure 6.9   Mission command in the C2 Approach Space.

Mission command has been categorized variously as a command technique, an overarching command philosophy, a coordination method to reduce complexity, a management method, a delegation method, a leadership model, and even a lifestyle.* Its notions of empowering the judgment and initiative of subordinates who are close to the action appeal to many modern theories of business and organizations, and thus it has recently attracted the attention of the nonmilitary business literature.†

6.7.3 Development of Mission Command in the Prussian and German Armies

Although approaches similar to mission command have probably been applied at times throughout history, modern notions of mission command have their roots in 18th and 19th century Prussia.‡ Because of the strong historical identification of mission command with the Prussian and German armies, many people refer to the concept by its modern German name of *Auftragstaktik*, meaning, literally, "order tactics." Because an order (*auftrag*) was meant to be general and to set up a mission without specifying every detail of execution, *Auftragstaktik* came to be translated imperfectly into English as "mission command." It is worth spending some time discussing the origins of the term, because it has now been so widely applied, and sometimes abused.

6.7.3.1 The Term "Auftragstaktik"   It is important to note that the term *Auftragstaktik* was *not* in general use in Prussia during most of the 19th century when the evolution of the concepts was taking place. In fact, the term appears not to have reached its widespread modern usage until after the creation of the *Bundeswehr* in 1956.

Over the course of the 19th century, the Prussian/German army came to distinguish between an order (*auftrag*) and a command (*befehl*). The first term had come to imply more discretion in execution than the second. The Field Regulations of 1888§ deliberately used the

---

* Wittmann (2012).
† Yardley and Kakabadse (2007); Howieson (2012); Bungay (2011).
‡ Leistenschneider (2002); Shamir (2011); Bungay (2005).
§ Prussia/Germany (1888a,b).

word *auftrag*, but the term *Auftragstaktik* does not appear in them as such. The term started to be used in the literature during the 1890s as debates took place within the German military between those who believed in nondetailed orders with discretion in execution (*auftrag*), and those who believed in more conventional rigid command relationships. Many other terms were used to denote mission command concepts between 1891 and 1914, including *Freies Verfahren* (free method), *Freie Taktik* (free tactics), *Auftragsverfahren* (order method), *Individualverfahren* (individual method), *Initiativeverfahren* (initiative method), and many others.[*] Citino[†] points out that as late as 1931, a German article discussing issues associated with centralizing or decentralizing command never uses the term *Auftragstaktik*.

6.7.3.2 Frederick the Great   The antecedents of Prussian mission command are to be found as far back as Frederick the Great.[‡] Frederick the Great's Prussian Army was one of the most effective in 18th century Europe. Some modern observers characterize Frederick's C2 as highly centralized and strictly hierarchical, with Frederick's brilliant guidance the key component. Certainly there is truth in this, but it does not quite tell the whole story. In fact, Frederick the Great planted some of the seeds that would later grow into mission command. Frederick believed that his subordinates, at least the ones at the highest echelons, needed to show initiative and pursue independent aggressive action to some extent.

Jorg Muth[§] recounts an anecdote from the Battle of Zorndorf (August 25, 1758) during the Seven Years' War. The battle was going badly for the Prussian side, and Frederick himself had to ride and capture a broken flag to rally a regiment in trouble. Meanwhile, there was a 50-squadron cavalry led by the young general Seydlitz, which had not yet entered the fray. Frederick sent an increasingly urgent series of orders for Seydlitz to attack, but Seydlitz disobeyed, thinking the time not yet right. When Seydlitz did finally attack on his own initiative, he achieved good results. Frederick did not excoriate

---

[*] Leistenschneider (2002).
[†] Citino (2004).
[‡] Friedrich II, 1712–1786.
[§] Muth (2011).

or punish Seydlitz, but instead declared that he had acted correctly. Nevertheless, Frederick was fighting during a time when it was still feasible for a single, brilliant commander to develop a synoptic view of the battlefield and exercise overall direction on site.

6.7.3.3 Prussian Defeat by Napoleon and Resulting Reforms By the beginning of the 19th century, the Prussian Army had retained its centralized and strictly hierarchical structure, but it no longer had Frederick the Great at its head, and it had lost whatever capacity it might have once had for individual initiative. It suffered a humiliating defeat at the hands of the French under Napoleon Bonaparte[*] at the battle of Jena-Auerstaedt in 1806.

Napoleon also ran a fairly centralized command, but there were several new elements. He employed open order tactics, with small, highly dispersed units compared with the more rigid formations of the time. Napoleon also employed relatively independent smaller armies called "corps." The corps had some independence of action by the standards of the day. Each corps was led by a Marshall, picked by Napoleon himself, with whom Napoleon could establish shared intent very effectively. Napoleon's army brought speed and chaos to the battle, confounding the slow and cumbersome Prussian command structure. Prussian soldiers were, overall, considerably better trained than Napoleon's peasant conscripts, and Prussian officers were brave. However, Prussian commanders simply would not make necessary decisions without permission from above, and this was an important factor contributing to their defeat. Of course, it also did not help the Prussians that their adversaries happened to be led by a military genius, whom Van Creveld characterized as "the most competent human being who has ever lived."[†]

The defeats led to an intensive review of the Prussian army and its practices, led by Gerhard von Scharnhorst.[‡] The resulting Act of 1809 restructured the army and established a general staff. The reformers also began to formalize the notions of independent initiative among

---

[*] 1769–1821.

[†] Van Creveld (1985), p. 64, as quoted by Bungay (2005).

[‡] 1755–1813.

commanders. The revised field regulations of 1812 contained some of this spirit.[*]

6.7.3.4 Von Moltke and the Development of Mission Command   Despite the reforms, change was not instantaneous, and old practices continued well into the middle of the 19th century. Field Marshall Helmuth Carl Bernhard Graf von Moltke[†] (known as "von Moltke the Elder," to distinguish him from his nephew, the German World War I commander[‡]) did perhaps more than anyone to bring notions of mission command to maturity. During the Prussian Army exercise maneuvers of 1858, he observed with disdain the sluggishness of the chain of command, and the lack of initiative displayed by officers. He clearly stated the need for orders to establish general intent, leaving commanders considerable latitude in their execution. He considered this necessary for the speed and flexibility of action necessary to confront the chaos and uncertainty of war. He put his concepts of general orders and commander initiative into practice to some degree in both the Austro–Prussian War of 1866 and the Franco–Prussian War of 1870 to 1871, both of which were major Prussian victories.

6.7.3.5 Austro–Prussian War   The exact extent to which the Prussians practiced mission command during the Austro–Prussian War of 1866 is debatable; not all von Moltke's commanders fully understood or agreed with the concept. It is important to remember that the idea and its practice were still evolving. Viewed in this light, we can identify important elements of mission command in the war. Von Moltke did issue fairly general orders, and did allow his subordinate commanders considerable initiative. The Prussian forces were dispersed and agile, using successive onslaughts of small units in seemingly chaotic patterns to confuse and surround the Austrians. Bungay[§] also points out an interesting fact illustrating von Moltke's thinking. In analyzing his victory over Austria at the Battle of Königgrätz (or Sadowa, July

---

[*] Leistenschneider (2002).
[†] 1800–1891.
[‡] Helmuth Johann Ludwig von Moltke, or "von Moltke the Younger," 1848–1916.
[§] Bungay (2005).

Figure 6.10 Field Marshall Helmuth Carl Bernhard Graf von Moltke ("von Moltke the Elder"). (From   http://commons.wikimedia.org/wiki/File:Helmuth_Karl_Bernhard_von_Moltke.jpg.   With permission.)

3, 1866), von Moltke remarked that two Austrian generals had taken independent action in defiance of their superior, Ludwig von Benedek, which had actually helped the Prussians win. However, he did not excoriate these generals. He believed their actions justified under the circumstances. This illustrates his belief that inaction is worse than sincere action that ultimately appears erroneous with hindsight. This would become an important cornerstone of German doctrine.

6.7.3.6 Franco–Prussian War   The French were impressed by Prussia's agility in the 1866 war with Austria, but they did not fully understand that some of that agility came from an evolution in command and control. They also saw weakness in the apparent fragmentation of the Prussian forces. The French began to emphasize defense over offense, and to favor relatively fixed positions and centralized control. Captain Ernst Schmedes, an Austro–Hungarian officer observing French maneuvers a year before the Franco–Prussian war, was shocked as opposing sides settled into fixed positions and refused to attack each other.*

While the Prussian conduct of the war was by no means perfect, and the Prussians made mistakes, the Prussians were definitely more agile and quicker to capitalize on opportunities provided by chance. They were able to encircle the French, hem them in, and ultimately win.

An interesting example of flexible command was the battle of Metz, on August 14, 1870. Bungay[†] presents a succinct and clear account of this incident, and we paraphrase his narrative closely. Von Moltke preferred to bypass Metz. However, he found himself in a position where French forces were massed there, and thought he would have to besiege it. Under von Moltke's orders, the Prussian First Army under General Steinmetz waited for the 2nd Army before investing the city. A brigade commander, Brigadier General Wilhelm von der Goltz,[‡] happened to observe some of the French forces retreating. A general retreat would allow the Prussians to bypass Metz at that time. There was no time to ask permission, so he attacked on his own initiative. As he got into trouble he was helped spontaneously by adjacent Brigadiers. When his commander, Steinmetz, found out, he was furious and ordered him to fall back. Von der Goltz essentially refused, sending back some token forces to appease his superior. Von der Goltz's actions had good results, but they were nevertheless highly controversial, with many observers finding them reckless and disobedient. However, in the end, he was praised in a 1910 German tactical field manual for taking appropriate spontaneous action.[§]

Perhaps nothing can better illustrate von Moltke's commitment to general orders and individual initiative than his general directive of August 30, 1870, to the Prussian Army, over a month after the Franco–Prussian war had started. Bungay's[¶] presentation and analysis of the directive is very instructive. The directive was only eight sentences long. Yet, in those eight remarkable sentences, von Moltke clearly summarized what was known about the situation, and what was not known; his overall intent; the role of each commander; and

---

* Wawro (2005).
† Bungay (2005).
‡ Wilhelm Leopold Colmar Freiherr von der Goltz (1843–1916).
§ Bungay (2005).
¶ Bungay (2011).

what to do—in a general sense—in case of an important contingency. The directive uses language such as, "it would be advisable, if at all possible" (when translated into English, of course), making it clear that commanders had freedom of action.

6.7.3.7 Drill Regulations of 1888   Debate about the merits of mission command continued after the Franco–Prussian war, with many naysayers. By 1888 von Moltke's essential ideas, championed by supporters such as General Sigismund Schlichting,[*] gained more formal doctrinal acceptance. The spirit of the 1888 *Exerzier-Reglement*, or drill regulations, for the infantry (see Figure 6.11),[†] embodied the importance of individual thought and action for commanders. The regulations state that higher commanders should "refrain from ordering more than they could or should order," and those commanders executing orders should "abstain from abusing the independence accorded to them." The regulations then state, "The exercise of independence within these limits, is the foundation of great results in war," and "These considerations are applicable to the lowest rank of commanders."[‡]

The 1888 regulations did not end the debate between proponents of mission command and those who espoused more traditionally circumscribed doctrines (the *normaltaktikers*). Debate was intense in the 1890s and continued into the 20th century. The *normaltaktikers* brought up many valid concerns, including the possible loss of cohesion and coordination if subordinates exercised independent initiative.[§] We must remember that modern tactical communications technologies did not exist and could not be used yet for coordination.

German military thinkers studied both the Boer War of 1899 to 1902, and the Russo-Japanese War of 1904 to 1905. The Boer War offered support both to the *Auftragstaktikers* and the *Normaltaktikers*. It appeared to validate the effectiveness of free-form tactics. However, it also demonstrated the need for control of small units. The Russo-Japanese war was largely seen as validating *Auftragstaktik*; as the

---

[*] Sigismund Wilhelm Lorenz von Schlichting (1829–1909).

[†] Prussia/Germany (1888a,b).

[‡] Prussia/Germany (1888b), p. 20.

[§] See Echevarria (2000) for an excellent detailed discussion of the debates. See also Hughes (1986, 1995).

Exerzir-Reglement

für

die Infanterie.

Berlin 1888.
Ernst Siegfried Mittler und Sohn
Königliche Hofbuchhandlung
Kochstraße 68—70.

Figure 6.11 Cover page of the 1888 German Field Regulations.

Japanese had been trained in German doctrine and methods. The 1906 drill regulations for the German infantry and the 1909 regulations for the cavalry both reemphasized the importance of subordinate initiative.[*]

6.7.3.8 World War I As we have seen above, the emerging concepts of mission command worked reasonably well for the Prussians/Germans in the Austro–Prussian War of 1866 and the Franco-Prussian War of 1870 to 1871.[†] However, those wars involved smaller

---

[*] Echevarria (1996, 2000).
[†] Fuhrmann et al. (2005).

armies and much smaller fronts, which were easier to handle with more rudimentary communications capability. Although mission command principles initially served the Germans well in the east as they defeated the Russians at Tannenberg (August 23–30, 1914), they were not applied as effectively in the west.

In the German western offensive of 1914, the deviations from the original offensive Schlieffen Plan, caused at least in part by field commanders taking initiative and responding to unfolding tactical conditions,* created a need for command, control, and communications technological capability that simply did not exist in 1914 to coordinate the movements of huge armies. The field commanders were taking actions without full knowledge of what other commanders were doing (they lacked shared awareness); without the immediate knowledge of headquarters; and crucially, without vital information that was often known to headquarters and other commanders (a lack of shared information). This ultimately contributed to the failure of the offensive because the conditions necessary for successful mission command were not established. We discuss this case in more detail in Chapter 7. As World War I evolved into static and bloody trench warfare, mission command principles became less relevant. The principles did help enable some temporary German successes in the later offensives of 1918.

6.7.3.9 Between the Wars   Between the two world wars, the Germans continued to refine and sharpen the principles of mission command, and to extend them further down the chain of command. Under the leadership of General Hans von Seeckt,† they intensively studied the lessons of World War I. Among the conclusions was a reaffirmation of the importance of general directives and individual initiative to help overcome the friction and chaos of battle.‡ Seeckt's ideas helped form an important basis of the 1933 *Truppenführung* (troop leadership).§ Echoing von Moltke, the *Truppenführung* rejected the idea that war

---

* For example, General von Kluck's turn to the Southeast away from Paris on August 31, 1914, and Prince Rupprecht of Bavaria's counterattack in Lorraine on August 8, 1914. See Fuhrmann et al. (2005).
† Johannes Friedrich "Hans" von Seeckt (1866–1936).
‡ English (1998).
§ Murray and Millett (2000).

could be centrally and meticulously planned, stating, "Situations in war are of unlimited variety. They change often and suddenly and only rarely are from the first discernable. Incalculable elements are often of great influence. The independent will of the enemy is pitted against ours. Friction and mistakes are everyday occurrences."* The *Truppenführung* emphasized mission command and independent initiative to overcome this chaos: "A commander must give his subordinates a free hand in execution so far as it does not endanger his objective."†

6.7.3.10 World War II   During World War II, the Germans applied the principles of mission command, along with fast, mechanized maneuver warfare, to devastating effect. Examples of mission command and individual initiative abound. We illustrate with one anecdote, recounted by Echevarria.‡ During the 1940 German offensive in the west, Hitler gave an order on May 24th to halt at the Aa River in Flanders. General Heinz Guderian,§ arriving at the scene a day later, found German forces had already crossed the river, in clear defiance. They did so on the individual initiative of Josef "Sepp" Dietrich, the division commander. Dietrich had judged that his position was vulnerable to potential enemy surveillance and attack from a well-placed hill called Mont Watten, and thus decided to take that high ground for his forces. Guderian disobeyed the order himself and crossed the river to find Dietrich. Far from reprimanding Dietrich, Guderian approved his disobedience, and ordered some of the 2nd Panzer Division to move up in support.

This example of mission command occurred within a larger context of top-down control by Hitler, who refused to lift the halt order despite the entreaties of some of his top commanders. In fact, while mission command in the World War II German forces was often very effective, it co-existed somewhat uneasily with Adolf Hitler's autocratic leadership. As the war progressed, Hitler increasingly tended to interfere with the armed forces' conduct of the war, especially in the

---

* Murray and Millett (2000).
† Murray and Millett (2000), p. 23; Bungay (2005).
‡ Echevarria (1986), pp. 51–52.
§ Heinz Wilhelm Guderian (1888–1954).

eastern front after 1942. It would be stretching the point much too far to say that this is why the Germans lost the war, but effective mission command was certainly a significant German asset that Hitler did not allow to flourish to full effect.

However repellent the ideologies propelling the Germans into war, and notwithstanding their ultimate loss, few would disagree that the fighting performance of the German armed forces was often exceptional. U.S. Army Colonel and military historian Trevor Dupuy estimated that the Germans consistently inflicted 50% more casualties per fighting soldier than did the British or Americans in World War II.[*] While his precise number has been challenged, the underlying truth of German effectiveness has been widely accepted.

The victors of both world wars gradually became convinced of the merits of mission command, leading to widespread attempts to adopt it. These have met with varying degrees of success, as outlined in detail by Eitan Shamir in his book[†] on the topic, and as we discuss briefly below.

6.7.4 Naval Traditions Resembling Mission Command

While they did not always use the terms "mission command" or *Auftragstaktik*, the navies of both Britain and the United States have a relatively long tradition of following, and sometimes enunciating, similar principles. In Chapter 1 we already discussed the approach, similar to mission command, used by Nelson at the Battle of Trafalgar in 1805.

The U.S. Navy also adopted important principles of subordinates' initiative. These principles were articulated clearly by the influential thinker Dudley Knox, in the period between 1913 and 1915,[‡] and they were embraced in World War II by Admiral Ernest King,[§]

---

[*] Dupuy (1976).
[†] Shamir (2011).
[‡] Knox (1913, 1914a,b, 1915).
[§] Ernest Joseph King (1878–1956).

Commander in Chief of the Atlantic Command. In a famous memorandum, King stated:

> I have been concerned for many years over the increasing tendency—now grown almost to "standard practice"—of flag officers and other group commanders to issue orders and instructions in which their subordinates are told "how" as well as "what" to do to such an extent and in such detail that the "Custom of the service" has virtually become the antithesis of that essential element of command—initiative of the subordinate.[*]

And

> It is essential to extend the knowledge and the practice of "initiative of the subordinate" in principle and in application until they are universal in the exercise of command throughout all the echelons of command.[†]

The principle of subordinate initiative was also clearly stated in the 1944 War Instructions for the U.S. Navy, also signed by King.[‡] The Navy did not employ mission command always or everywhere in World War II, but in many instances it did, usually with success. One example was the initiative shown by Captain Arleigh Burke[§] and his subordinates at the Battle of Cape St. George (November 25, 1943), which Admiral William F. Halsey[¶] dubbed the "Trafalgar of the Pacific."[**]

6.7.5 Modern Doctrinal Statements Supporting Mission Command

In the United States, Army publication FM 6-0 "establishes mission command as the C2 concept for the Army."[††] The publication further states:

> Mission command relies on subordinates effecting necessary coordination without orders. While mission command stresses exercising subordinates' initiative at the lowest possible level, all soldiers recognize that doing so may reduce synchronization of the operation. Thus,

---

[*] King (1941), p. 1.
[†] Ibid.
[‡] King (1944).
[§] Arleigh Albert Burke (1901–1996).
[¶] 1882–1959.
[**] Palmer (2005).
[††]U.S. Army (2003), pp. viii, 2–22.

commanders accept the uncertainty that accompanies subordinates exercising initiative. Their trust in subordinates they have trained gives them the assurance that those subordinates will direct actions that will accomplish the mission within the commander's intent.

The U.S. Marine Corps has also established mission command as its official doctrine:

> The Marine Corps' concept of command and control is based on accepting uncertainty as an undeniable fact and being able to operate effectively despite it. The Marine Corps' command and control system is thus built around mission command and control which allows us to create tempo, flexibility, and the ability to exploit opportunities but which also requires us to decentralize and rely on low-level initiative.[*]
>
> Mission command and control tends to be decentralized, informal, and flexible. Orders and plans are as brief and simple as possible, relying on subordinates to effect the necessary coordination and on the human capacity for implicit communication—mutual understanding with minimal information exchange. By decentralizing decision-making authority, mission command and control seeks to increase tempo and improve the ability to deal with fluid and disorderly situations.[†]

The U.S. Marines even go so far as to caution against procuring any equipment that would enable an interference with mission command:

> Equipment that permits over control of units in battle is in conflict with the Marine Corps's philosophy and is not justifiable.[‡]

When officers from the First Battalion, 2nd Marines (a unit with extensive combat experience in Iraq) were interviewed by researchers from the Naval Postgraduate School about operational and technological needs, they had a "hesitancy to embrace any technology that would allow those higher echelons the temptation of micro-managing small unit actions from behind a plasma screen."[§]

---

[*] USMC (1996), p. 104.
[†] Ibid., p. 79.
[‡] USMC (1989), p. 52.
[§] Senn and Turner (2008), p. 13.

The U.S. Air Force observes that a "reluctance to delegate decisions to subordinate commanders slows down C2 operations and takes away the subordinates' initiative."*

General Martin Dempsey,† the Chairman of the U.S. Joint Chiefs of Staff, published a "White Paper" on mission command in 2012.‡ In it, he articulated the importance of shared context and understanding, observing that

> In its highest state, shared context and understanding is implicit and intuitive between hierarchical and lateral echelons of command, enabling decentralized and distributed formations to perform as if they were centrally coordinated. When achieved, these practices result in decentralized formal decision-making throughout the force, leading implicitly to the opportunity to gain advantageous operational tempo over adversaries.§

General Dempsey ends his paper with

> *Understand my intent*: I challenge every leader in the Joint Force to be a living example of Mission Command. *You have my trust.*¶

In the United Kingdom, mission command is a stated cornerstone of defense policy**:

> The United Kingdom's philosophy of command is based on mission command, which promotes initiative, decentralised [sic] command, and freedom and speed of action, yet remains responsive to superior direction.††

The United Kingdom also makes an explicit connection between mission command and net-enabled capability (NEC):

> [I]ncreasing degrees of NEC permit mission command to be extended, with confidence, down through the tiers of command.‡‡

---

* USAF (2007), p. 11.
† Born 1952; became Chairman of the Joint Chiefs of Staff in 2011.
‡ Dempsey (2012).
§ Ibid., p. 4.
¶ Ibid., p. 8.
** UK Army (2005); UKMoD (1989); UKRAF (2008).
†† UKRAF (2008), p. 61.
‡‡ Ibid., p. 65.

Mission command is also important in Canadian doctrine,* Dutch doctrine,† and many others.

6.7.6 Doctrine Is Not Enough

The adoption of a doctrine embracing mission command and tenets of decentralized C2 may be necessary for those militaries that wish to implement such concepts, but it is not sufficient. Moving in an agile manner from a centralized, purely hierarchical C2 paradigm to a more decentralized one, even in only a subset of situations, requires considerable effort in changing the command culture. Higher levels of command must become accustomed to delegating and not overspecifying or micromanaging missions. Lower levels must become accustomed to taking initiative and not receiving highly detailed orders. The Prussian and German doctrine and its application took over a century to evolve, allowing a thorough inculcation of the mission command tenets of trust and simple orders into the military culture.‡

Thus, simply "cutting and pasting" mission command concepts into doctrine is unlikely to yield successful results.§ It has been observed, for example, that the British Army in World War II was theoretically working from a fairly decentralized doctrine but did not behave accordingly and did not reap the associated benefits.¶ Stewart (2006) quotes other observers who noted that the Germans and Italians had similar doctrine in that war but achieved very different levels of success, probably owing to military cultural factors. Stewart (2006) discusses a number of examples of military organizations with mission-command doctrines behaving in a hierarchical and centralized fashion, depending in part on the personalities of individual commanders. Stewart (2009) summarizes a number of studies showing significant variations in the importance accorded to understanding command intent—an essential aspect of mission

---

* Canada DND (1996).
† Vogelaar and Hanskramer (2004).
‡ Wyly (1991).
§ Oliveiro (1998).
¶ Johnston (2000).

command—by officers of doctrinally decentralized western military organizations.

Even in a military organization with a mission command doctrine, commanders with a tendency to micromanage may do so if the general culture allows it. The same technologies that facilitate decentralization and the transmission of intent can sometimes also facilitate micromanagement. Zagurski (2004) notes that in some early U.S. Army experiments with fully digitized brigade combat teams, some commanders attempted to micromanage. In its Fleet Battle Experiments-India (FBE-I) of 2001, the U.S. Navy tested decentralized execution of joint fires. The experiments showed that operational commanders often have considerable difficulty allowing decentralized execution.* In the real world, the pressure to micromanage can be significant, given the visibility of modern tactical operations to upper command echelons and the media.†

These considerations underscore the importance of education and the development of individuals to the successful adoption of mission command or any flavor of net-enabled decentralized C2. They also raise the issue of how difficult cultural change may be, if the educators themselves are imbued with the existing culture.

6.8 General Remarks

From the discussion above, we may glean a number of suppositions. One is that for rapidly changing, chaotic environments and problem spaces that are not neatly defined and decomposable, some level of decentralization in C2 may be effective. This decentralization will, most importantly, involve a broader allocation of decision rights, but will also very likely involve less-constrained patterns of interaction and broader distribution of information than the classic, industrial-age, strictly hierarchical version of C2.

Traditional hierarchical organizations may be well adapted to some problems and circumstances but not to others. In particular, hierarchies may be well adapted to stable, decomposable problems, where parts of the problem space can be parceled out and solved more or less

---

* Saunders (2002).
† Fox (1995); Ramshaw (2007).

independently. However, they may not be as well adapted to problems that are complex and dynamic (Big Problems).

It is important for enterprises to be able to choose the most appropriate approach for their missions and circumstances.

In Chapter 8, we provide some experimental tests of the above suppositions, and several related ones.

7

How C2 Goes Wrong

7.1 Introduction

In this chapter, we seek to understand how Command and Control (C2) can go wrong by studying several historical situations that have been characterized as experiencing one or more C2 failures of varying severity.* Our ultimate purpose is to understand better how enterprises can minimize the chance of such failures. Many of the cases we study here predate the emergence of most of the megatrends now reshaping the C2 landscape, so it does not make sense to view them specifically in that light. The exception is that almost all of them qualify as "Big Problems" using the considerations of Chapter 3.

By "C2 failure," we mean an observed inability to carry out adequately the functions associated with C2. In some of the 20 cases presented here, the C2 failure(s) had an obvious detrimental effect on the mission in question. In others the operational effect was more difficult to discern or was overwhelmed by other factors. Thus, not all of the cases where C2 failures were observed also involved mission failure. In fact, in many cases, the overall mission was judged a success despite these C2 failures. However, in all cases, the failure of C2 had some detrimental impact on the mission.

The situations under discussion and some basic data about them are listed in Table 7.1. Some involve military operations, including ones from World War I and World War II, Operation Desert Storm, the U.S. Iranian Hostage Rescue mission of 1980 (discussed in Chapter 1), the Russia-Georgia war of 2008, and others. Others involve the run-up, and response to, terrorist attacks such as those of September 11, 2001. Still others revolve around the responses to major disasters,

---

* In all of the cases considered here, the literature describing what happened explicitly refers to "command and control failure." See also Vassiliou and Alberts (2013).

Table 7.1   Situations Discussed: Basic Data

INCIDENT	WHEN	WHERE	RESULT	NOTES
MILITARY OPERATIONS				
Great Retreat of 1914, First World War	August 24–31, 1914	France	Unraveling of the British Cavalry Division	[1]
Run-Up to First Battle of the Marne, First World War	August and September 1914	France	Allied victory in the ensuing battle on September 5–12, 1914; 550,000 estimated total casualties in ensuing battle	[2]
First Battle of Savo Island, Guadalcanal Campaign, Second World War	August 8–9, 1942	Savo Island	Allied defeat; 1135 combat deaths, of which 1077 allied	[3]
Mayaguez Incident	May 12–15, 1975	Koh Tang Island, Gulf of Siam	Mission success for United States after heavy fighting; c. 88 dead, 105 injured (combatants, both sides)	[4]
U.S. Hostage Rescue Mission	April 24–25, 1980	Iran	Mission failure; 8 U.S. servicemen dead.	[5]
U.S. Invasion of Grenada	October 25, 1983	Grenada	Mission success for United States; 89 deaths, 533 injuries (combatants, all sides)	[6]
First Gulf War, Operation Desert Storm	January 17–February 28, 1991	Iraq	Coalition victory; 358 coalition combat and theater deaths; c. 22,000 Iraqi military deaths	[7]
Russia–Georgia War	August 7–16, 2008	Georgia, Abkhazia, South Ossetia	Russian victory; estimates of total combat deaths (all sides) vary from c. 500–3200	[8]

INCIDENT	WHEN	WHERE	RESULT	NOTES
TERRORIST ATTACKS				
Oklahoma City Bombing	April 19, 1995	United States (Oklahoma)	168 deaths, 680 injuries	[9]
9/11 Attacks	September 11, 2001	United States (New York, Virginia)	2996 deaths, over 6000 est. injuries, $40 billion in insurance claims, unquantifiable total impacts	[10]

	WHEN	WHERE	RESULT	NOTES
7/7 London Bombings	July 7, 2005	United Kingdom (London, England)	56 deaths, c. 700 injuries	[11]
2011 Norway Attacks	July 22, 2011	Norway (Oslo and Utøya Island)	77 deaths, 319 injuries	[12]

DISASTERS AND EMERGENCIES

INCIDENT	WHEN	WHERE	RESULT	NOTES
King's Cross Underground Fire	November 18, 1987	United Kingdom (London, England)	31 deaths, 100 estimated injuries	[13]
Clapham Railway Junction Accident	December 12, 1988	United Kingdom (London, England)	35 deaths, 500 estimated injuries	[14]
Hillsborough Stadium Disaster	April 15, 1989	United Kingdom (Sheffield, England)	96 deaths	[15]
Hurricane Andrew	August 24, 1992	United States (Florida)	26 direct deaths, 39 indirect; $20 billion estimated damages	[16]
Columbine High School Shootings	April 20, 1999	United States (Colorado)	13 deaths, 24 injuries	[17]
Indian Ocean Tsunami	December 26, 2004	Indonesia, India, Sri Lanka, Thailand, Maldives	227,898 deaths, c. 1.7 million people displaced	[18]
Hurricane Katrina	August 23–30, 2005	United States (Louisiana, Mississippi, and environs)	1836 deaths, $75 billion damages, $110 billion total economic impact	[19]
Black Saturday Fires	February 7, 2009	Australia (Victoria)	173 deaths, AUS$4 billion in economic impact	[20]

Notes: [1] Gardner (2009). [2] Fuhrmann et al. (2005); Winter (2010). [3] Frank (1990). [4] U.S. Government Accountability Office (1976). [5] Holloway (1980); Bowden (2006). [6] Cole (1997). [7] Tucker (2010). [8] Independent International Fact-Finding Mission on the Conflict in Georgia (2009). [9] Shariat et al. (1998); Oklahoma Department of Civil Emergency Management (2003). [10] Makinen (2002). [11] Lieberman and Cheloukhine (2009). [12] Commission on 22 July (2012). [13] Fennell (1988); Croome and Jackson (1993). [14] Hidden (1989). [15] Hillsborough Independent Panel (2012). [16] Rappaport (2005). [17] U.S. Fire Administration (1999). [18] USGS (2013). [19] Knabb et al. (2005). [20] Parliament of Victoria, 2009 Victorian Bushfires Royal Commission (2010).

such as Hurricane Katrina, or responses to smaller emergencies, such as London's King's Cross Underground Fires of 1989. The list is by no means exhaustive, nor is it a uniform sample of all C2 failures in the past hundred years. Rather, it is an illustrative collection of notable incidents involving a recognized failure in C2.

7.2 C2 Failures

C2 failures generally manifest themselves as a lack of access to information or an absent, incomplete, irrelevant, delayed, or erroneous transfer of information from those who have it to those who need it. This lack of information quality may be caused by failure of the information infrastructure, or "infostructure," to satisfy the requirements of the mission or circumstances. It may also be caused by behavioral failures. These causes may in turn be a result of preexisting or *a priori* problems, or current stresses with which individuals, organizations and/or systems cannot adequately cope. Some C2 failures can thus be traced back to poorly designed organizations or sometimes inherently good organizational design that is maladapted for a particular mission or circumstance.

We make a distinction between an *inability* to communicate information and a *failure* to do so, in a timely manner, when it would have been appropriate. An *inability* to communicate information may be further subdivided into an *inability caused by system design and policy shortfalls*, and an *inability caused by circumstance*. Examples of the former are interoperability problems, a preventable shortage of equipment or bandwidth, or self-imposed security constraints. Examples of the latter include physical constraints, destruction of infrastructure or equipment, or denial by adversary (although this last might also sometimes be classified as a system design shortfall). A *failure* to communicate, when the means to do so are available, may result from human error, organizational silos, or mistrust. A failure to communicate may occur when the technical ability to communicate is present but the knowledge of how to communicate is missing, sufficient incentives are not present, or individuals and organizations are unwilling to communicate with one another.

A lack of communication, whatever the cause, can have an adverse impact on operations, possibly resulting in mission failure. Even if the mission ultimately succeeds, C2 failure(s) may cause missed opportunities, duplication of effort, delays, and reduced effectiveness. This can happen even if the organization design is sound and appropriate, although good organization design may make the system more resilient to communication problems. Conversely, an inappropriate organization or C2 Approach does not guarantee a failure to communicate, but it makes such failures more likely. It also makes a lack of communication potentially more serious.

The inability to communicate may often be addressable with continuing investments in research, development, technology, and engineering (RDT&E), in information and communications technology (ICT), as well as smart and effective acquisition policies for communications equipment and ICT. Behavioral failures to communicate are addressable with training, doctrine, and proper organization design, as well as agile C2 Approaches.

7.3 *A Priori* Structural Problems

C2 failures often occur because of a predisposition to such failures that is inherent in the organization and the systems that support C2. This was the case for many of the situations and incidents.

For example, in the run-up to the attacks of September 11, 2001 (9/11), the organizations responsible for military air defense and those responsible for the management of civil air traffic each had their own independent hierarchical structures and silos that promoted vertical communication. This lack of effective cross-coordination mechanisms resulted in sufficient delays that there was not enough time to shoot down the hijacked planes that successfully collided with the World Trade Center in New York and the Pentagon in northern Virginia.[*]

In the immediate response to the 9/11 attacks, significant organizational seams between the New York Police Department (NYPD), the Fire Department of New York (FDNY), and the Port Authority Police Department (PAPD) proved problematic. These seams, exacerbated by the communications difficulties discussed below, resulted in redundant searches for civilians and other instances of inefficient resource deployment.[†]

---

[*] Grant (2006).
[†] National Commission on Terrorist Attacks upon the United States (2004).

The failed Iran Hostage rescue attempt, discussed in Chapter 1, involved several different entities that suffered from compartmentalization and evidenced mutual distrust between them. These seams, combined with the communications problems described in Chapter 1, and a good bit of bad luck, led to mission failure and eight dead U.S. servicemen.

Similarly, the U.S. service components involved in the *Mayaguez* response of 1976 were not organized to form a fully cohesive task force. The planning process was disjointed, as described further below, and there was insufficient unity of effort between the U.S. Marines, Air Force, and Navy.* There were also shortages of communication equipment on the ground, as discussed in Section 7.4. On top of all this, there was unfortunate micromanagement and interference from the upper echelons in Washington, DC. At one point in the heat of battle, Marines had to respond to an information request from Washington: did they have a Khmer interpreter with them?† Although the United States achieved victory in the Battle of Koh Tang and recovered the *Mayaguez* and ultimately its crew, the fighting was very difficult and the margin of victory was small. C2 problems were ultimately overcome by initiative and heroism.‡

The Russian armed forces, although they won their war with Georgia fairly quickly in 2008, arguably had too difficult a time doing so and suffered too many casualties. C2 failures involving coordination and communication were part of the problem, as discussed in Section 7.5. Here we observe that the Russian military was organized along Cold War and even World War II principles, for large fights involving massive armies. There was little overall coordination between the Army, Air Force, and Navy suitable for joint prosecution of a relatively small operation. One retired general argued that the subordination of Army aviation to the Russian Air Force (*Voyenno-Vozdushnyye Sily*—VVS) was at the core of the failures to provide close air support to ground combat forces, and called for the return of such tactical aviation to the control of Army ground units.§ It is worth noting that the Georgia war served as a stimulus for reform of

---

* Toal (1998).
† Ibid.
‡ Ibid.
§ McDermott (2009).

the Russian military, away from its Soviet legacy and toward better performance in smaller, faster operations.*

The "Great Retreat" of the British forces in 1914, in the face of an initial German offensive, offers another example of an inappropriate approach to C2. The heads of cavalry brigades were used to taking initiative and not being micromanaged. This served the British well in the various "small wars" of the world-spanning empire, but created problems in the huge conflict that was World War I. Thus, while retreating cavalry brigades had difficulties in communicating with General Allenby† at his headquarters, this was not the only cause of the communication deficit, since they were not terribly inclined to communicate in the first place. As the brigades retreated, they completely lost touch with each other and headquarters, and the British Cavalry effectively disintegrated as a viable unit for a time. This had significant adverse consequences on the ability to accomplish the mission. In the battle of Le Cateau on August 26, Allenby was unable to offer any assistance to the Second Army Corps—his brigades were effectively gone.

During Operation Desert Storm, there was a lack of coordination between the U.S. Army and the U.S. Air Force over the crucial placement of the Fire Service Control Line (FSCL).‡ The FSCL is the boundary within which all air strikes must be coordinated with ground commanders to avoid fratricide, and outside of which air strikes are more freely conducted. The Army tended to want to position the line relatively far out, to give it more room to operate. The problem was that if the Army did not move fast enough to justify such a placement, the line was typically not repositioned. This hampered the Air Force from pursuing fleeing enemy forces and served to shield significant portions of the Iraqi Army. As one observer put it, "the safest place for an Iraqi to be was just behind the FSCL."§

Turning now to disaster response, organizational deficits have been a persistent problem in responses to major disasters (Figure 7.1).¶ In the response to the Indian Ocean Tsunami of 2004, there were

---

* Cohen and Hamilton (2011).
† Edmund Henry Hynman Allenby, First Viscount Allenby (1861–1936).
‡ McDaniel (2001).
§ Ibid., p. 2.
¶ Donahue and Tuohy (2006).

militaries from 11 countries involved. Each had a somewhat different relationship with the Indonesian Government. A case study undertaken in support of NATO SAS-065* noted many examples of a lack of shared intent.† There was a lack of coordination between the various military establishments involved, between military establishments and international nongovernmental organizations (NGOs), between international NGOs and Indonesian NGOs, and between U.S. and UN agencies. A Humanitarian Information Center was established in an effort to provide some oversight for hundreds of NGOs. Their daily meetings were characterized as being "unwieldy" and as "a shambles."‡

In the response to Hurricane Katrina, the roles of various U.S. federal agencies were not properly delineated. There was overlap and conflict among them, and between them and the states of Louisiana and Mississippi, as well as local agencies and other actors.§ As an example, both local police and the National Guard were working at the Louisiana Superdome, which served as an evacuation center, but each side said the other was supposed to lead. This led to security problems, and many responders left.¶ The House of Representatives report on Katrina** specifically mentions structural *a priori* coordination deficits between the Department of Defense (DoD) and the Federal Emergency Management Agency (FEMA), and between both of them and the State of Louisiana. Similar observations about the lack of proper role delineation and interagency coordination can be found in the reports on the responses to other disasters and emergencies. During Australia's Black Saturday fires of 2009,†† the State of Victoria's Country Fire Authority (CFA) and Department of Sustainability and the Environment (DSE) reportedly followed distinct and inconsistent operating procedures. In the response to London's King's Cross

---

* NATO System and Analysis Studies group 065: Network Enabled Capability (NNEC) C2 Maturity Model.
† Huber et al. (2008).
‡ Ibid., p. 4.
§ Moynihan (2006); U.S. House of Representatives (2006); U.S. Senate (2006).
¶ Moynihan (2006).
** U.S. House of Representatives (2006).
†† Parliament of Victoria, 2009 Victorian Bushfires Royal Commission (2010).

(b)

(a)

Figure 7.1 (a) U.S. soldiers on patrol on Canal Street in New Orleans, Louisiana, in the aftermath of Hurricane Katrina, September 5, 2005. (From the U.S. Department of Defense, http://www.defenselink.mil/photos/Sep2005/050904-A-7377C-005.jpg. With permission.) (b) Japan Ground Self-Defense Force in relief operations after the 2004 Indian Ocean Tsunami. (From the Japan Ground Self-Defense Force, https://picasaweb.google.com/lh/photo/lySIZSL8VWdNNcEEpNuC7NMTjNZETYmyPJy0IiipFm0. With permission.)

Underground fire in 1987,* there was poor coordination between the London Underground and police and fire agencies.

It is worth noting that even relatively decentralized and supposedly "nimble" organizations, such as terrorist groups, are not always immune to organizational seams and structural problems. A study of the decline of Al Qaeda in Iraq recounts that cells often became bureaucratic and compartmentalized, with a military officer, security officer, *sharia* officer, and administrative officer. One individual was quoted as saying that "there is little cooperation between the four elements."†

In many observations of the situations and incidents discussed here, there is a recurring refrain that "no one was in charge." Table 7.2 compiles many quotes to this effect. To see them all in one place is quite striking. However, we must note that in complex endeavors, there may indeed *be* no one in charge. It may not be necessary or desirable to have a single organization "in charge,"‡ as long as the effort is properly coordinated, participants understand their roles and possess sufficient shared awareness, and communication channels are available if required. The refrain "no one is in charge" is convenient shorthand for a disorganized and uncoordinated effort.

7.4 Behavioral Failures to Communicate

Many problems in C2 arise from a failure to access available information, or to communicate necessary information to those who need it. We are not speaking here of an *inability* to communicate, as in Section 7.5. The *means* to communicate may exist, but communication does not take place for any number of reasons, including a lack of *will* or *incentives*. In this section we consider communication deficits caused by any number of organizational problems, such as poorly delineated roles and responsibilities, bureaucratic silos that ignore or even mistrust each other, or simple human error. The types of *a priori* structural problems discussed in Section 7.2 do not guarantee failures of communication,

---

* Fennell (1988).
† Fishman (2009), p. 20.
‡ Alberts et al. (2010).

Table 7.2 "No One in Charge"

INCIDENT	QUOTE	REFERENCE
Black Saturday Fires Response	"…roles of the most senior personnel were not clear, […] no single agency or individual in charge…."	Parliament of Victoria, 2009 Victorian Bushfires Royal Commission (2010), p. 8
Hurricane Andrew Response	"…failure to have a single person in charge with a clear chain of command."	Florida Governor's Disaster Planning and Response Review Committee (1992), p. 60
9/11 Attacks	"…no one was firmly in charge of managing the case…. Responsibility and accountability were diffuse." [about intelligence]	National Commission on Terrorist Attacks upon the United States (2004), p. 400
King's Cross fire response	"…uncertainty over which of the London Underground staff was in charge…. "	Fennell (1988), pp. 73–74
Iran hostage rescue	"…confusion about 'who was in charge.'"	Anno and Einspahr (1988), p. 10
	"…uncertainty as to who was in charge."	Thomas (1987), p. 10
	"…no one who was in overall charge."	Gass (1992), p. 15
	"…no way to quickly find out or locate who was in charge."	Holloway (1980), p. 51
Mayaguez incident response	"…[planning activity] lacked coordination…No one seemed to be in charge."	Toal (1998), p. 18
Hurricane Katrina response	"…no single individual who took charge…"; "State officials and FEMA disagreed about who was in charge."	Moynihan (2006), pp. 22, 24
	"Too often, because everybody was in charge, nobody was in charge."; "…no consensus on who was in charge."; "… disagreed on who was in charge, could not find out who was in charge, or did not know who was in charge."	U.S. House of Representatives (2006), pp. xi, 185, 186
Indian Ocean tsunami response	"…coordinating meetings were 'very unwieldy' and 'internal coordinating meetings were a shambles.'"	Huber et al. (2008), p. 4
Columbine High School shootings	"…Who's in Charge? No one could answer the question."	Moody (2010), p. 39

Source: From Vassiliou and Alberts (2013).

but they create the conditions making such failures more likely, particularly under the stress of a battle or a disaster response.

An example of a failure to communicate occurred during the Guadalcanal campaign of late 1942, during World War II. The Guadalcanal campaign is generally credited as a success for the Allied powers, although it involved some significant losses in individual battles, and the Allies did not achieve full dominance in the regional waters around Guadalcanal. One of the battles in the campaign that the Allies lost outright was the Battle of Savo Island (Figure 7.2) on August 8, 1942. The communication failure occurred in the cruiser groups constituting the Allied screening force guarding against a Japanese naval attack.* On the night of the battle, the commander of the screening force, Rear Admiral Victor Alexander Charles Crutchley,† took his ship out of the southern cruiser group in order to attend a conference with Admiral Richard Turner. However, he did not inform his second in command, Captain Frederick Riefkohl,‡ who was in the northern cruiser group. Riefkohl remained ignorant that he was now in command of the screening force. When the Japanese attacked, there was no coordinated response. Moreover, a crucial radio message warning of an impending attack was not relayed to Riefkohl because of human error. These C2 failures were not the only causes of the loss of the Battle of Savo Island, but they adversely impacted the mission.

Another example of a C2 failure not caused by technical means can be found in the planning of the U.S. response to the capture of the *Mayaguez*. Planning cells were physically separated from each other and did not exchange much information. Many staff members were also absent. As a result, there was not a unified and coordinated plan, and many participants remained unaware of whatever plans there were.§

Responses to natural disasters have been rife with failures to communicate. Consider Hurricane Katrina. Moynihan¶ gives many examples, but here we will recount only two for the sake of illustration. The Louisiana Superdome football stadium served as a collection center for

---

* Hone (2006).
† 1893–1986.
‡ Frederick Lois Riefkohl (1889–1969).
§ Toal (1998).
¶ Moynihan (2006).

Figure 7.2 View of Savo Island from the East, before the battle. (From the U.S. Defense Department, USONI, 1943. With permission.)

people who would later be evacuated. FEMA had an evacuation plan and was more or less ready to execute. The commander of Joint Task Force Katrina, General Russel L. Honoré, told the National Guard to cancel the plans—but he did not inform FEMA. This delayed the evacuation by at least a day. In another example, New Orleans mayor Ray Nagin declared the Ernest N. Morial Convention Center as a refuge but did not broadly communicate this decision. FEMA and the Department of Homeland Security (DHS) did not realize until two days later, when about 19,000 people were stranded at the convention center without supplies.

Similar failures have also occurred in the responses to lesser emergencies. In the 1989 Hillsborough Stadium incident, "communications between all emergency services were imprecise and inappropriately worded, leading to delay, misunderstanding, and a failure to deploy officers to take control and coordinate emergency response."* After the Oslo bombing of 2011 but before the mass shootings on the same day, a citizen gave police a description of the likely perpetrator, as well as his vehicle license number. The officers did not pass the information up the command chain for at least 20 minutes, and it did

---

* Hillsborough Independent Panel (2012), p. 12.

not reach the right people for two hours, by which time the shootings on Utøya Island had already begun.[*]

7.5 Inability to Communicate

There are many cases where it is simply not possible to access or share information, or for two or more entities to coordinate. An inability to communicate may result from many causes. We may group these causes into two categories. The first is an inability to communicate because of system design or policy shortfalls. These shortfalls may be a manifestation of the types of *a priori* structural problems discussed above, or they may arise from a different set of dysfunctions within the establishment. Examples of an inability to communicate because of system design or policy shortfalls include

- Lack of interoperability in communications equipment
- Preventable shortages of equipment or bandwidth
- Self-imposed security constraints

An inability to communicate may also arise because of circumstances. For example,

- Communication may be physically impossible with current technology or the systems available
- Infrastructure or equipment may be damaged or destroyed
- Communication may be denied by an adversary

7.5.1 Inability to Communicate Because of Circumstances

7.5.1.1 Destruction of Infrastructure or Equipment   Infrastructure failure is a common and serious problem during major disasters.[†] The disaster that creates the conditions demanding a response also destroys or incapacitates the communications infrastructure. This happened during Hurricane Andrew,[‡] Hurricane Katrina,[§] and the Indian Ocean

---

[*] Dennis (2012).
[†] Donahue and Tuohy (2006).
[‡] Florida Governor's Disaster Planning and Response Review Committee (1992).
[§] Moynihan (2006); U.S. House of Representatives (2006); U.S. Senate (2006).

Figure 7.3   U.S. Marines board the *Mayaguez*, May 15, 1975. (From the U.S. Defense Department, USMC A706262. http://www.history.navy.mil/photos/numeric/numbers.htm. With permission.)

Tsunami.* Even if not substantially destroyed, infrastructure may be overwhelmed by the communication demands imposed by the response to the emergency. This was a factor in the responses to the 9/11 attacks† and Australia's Black Saturday fires,‡ as well as the responses to the Oklahoma City bombing§ and the Hillsborough Stadium disaster.¶

In the *Mayaguez* incident of 1976 (Figure 7.3), a serious communications problem was precipitated by the destruction of ultrahigh-frequency (UHF) radios in a helicopter crash. The remaining very-high-frequency (VHF) radios were overloaded, making communication between aircraft and Marines on the ground very difficult. This resulted in crucial problems coordinating air strikes.**

7.5.1.2 Physical Impossibility During the British "Great Retreat" of 1914 in the beginning of World War I, communications collapsed completely as cavalry brigades separated from each other.

---

* Huber et al. (2008).

† National Commission on Terrorist Attacks upon the United States (2004).

‡ Parliament of Victoria, 2009 Victorian Bushfires Royal Commission (2010).

§ Oklahoma Department of Civil Emergency Management (2003).

¶ Hillsborough Independent Panel (2012).

** Toal (1998).

Communications depended greatly on motor vehicles, and the roads of northern France had become so clogged as to be almost impassable.* With the technology available at the time, communication was not physically possible.

In another example from the early stages of World War I, the German offensive of 1914 was hampered severely by coordination failures resulting in part from an inability to communicate.† This contributed to the German defeat in the First Battle of the Marne, which in turn dashed German hopes for a quick victory and also signaled the end of mobile conflict and the beginning of bloody trench warfare. We discuss this case in more detail later in the chapter.

7.5.1.3 Denial by Adversary Among the cases considered in this chapter, the German western offensive in 1914, discussed in more detail in Section 7.7.5, offers an example of this. The Germans found their wireless communications disrupted by a French jamming station atop the Eiffel Tower.‡

7.5.2 Inability to Communicate Because of System Design or Policy Shortfalls

7.5.2.1 Shortage of Appropriate Equipment or Bandwidth In almost all the incidents considered here, there was a shortage of appropriate communications equipment or bandwidth. The following are some examples.

During Russia's 2008 war with Georgia, the Russian forces did not have enough communications equipment, and what they did have was antiquated and often not interoperable (see Section 7.5.2.2). Commanders ended up relying on personal mobile phones for C2. Worse yet, the calls had to go through the enemy's infrastructure, as the South Ossetian cellular networks were run by Georgia. In one instance, the 58th Army Commander, Lieutenant Anatoliy Khrulev, reportedly had to borrow a satellite telephone from a journalist in order to communicate with his forces.§

---

* Gardner (2009).
† Fuhrmann et al. (2005).
‡ Ibid.
§ McDermott (2009).

In the 1983 U.S. Invasion of Grenada, besides the interoperability problems discussed below, there was also a shortage of satellite communications.[*]

During the response to the 9/11 attacks, FDNY radios performed very badly inside buildings. A repeater system that had been set up to solve such problems was not properly activated because of human error. A shortage of bandwidth also plagued both the NYPD and FDNY.[†] In an earlier instance of radios having problems working indoors, responders to the King's Cross Underground Fire in London had severe difficulties with radio communication underground.[‡] In the response to Hurricane Andrew, there was a severe shortage of high-frequency (HF) radios.[§]

7.5.2.2 Lack of Interoperability Even the most expensive and sophisticated communications equipment is of little use if it cannot talk to other communications equipment. A lack of interoperability between communications equipment or information technology systems was a common problem in many of the situations studied. Chapter 1 discusses how many of the entities involved in the Iran hostage rescue mission could not communicate with each other because of interoperability problems.

Another U.S. military example is the successful invasion of Grenada in 1983. During this invasion, Marines in the north and Army Rangers in the south used their radios in such a way that interoperability was impeded, and they could not talk to each other. When the Marines ran into trouble at one point, the Rangers did not know about it for an unacceptably long time. Interoperability problems also led to a highly publicized incident in which a soldier had to call for air support by placing a commercial long distance telephone call from Grenada to Fort Bragg, North Carolina.[¶]

---

[*] Anno and Einspahr (1988).
[†] National Commission on Terrorist Attacks upon the United States (2004).
[‡] Fennell (1988).
[§] Florida Governor's Disaster Planning and Response Review Committee (1992).
[¶] Anno and Einspahr (1988).

Interoperability problems plagued the Russians during their war with Georgia in 2008. Ground units were unable to communicate with space-based and electronic intelligence assets. As a result, the Russians could not employ their electronic warfare systems to full advantage to suppress Georgian air defenses and could not make full and effective use of satellite targeting support or precision guided munitions. There were also interoperability problems between the units of different services of the Russian armed forces. Ground commanders had very little control over needed air support. Reportedly, Colonel General Aleksandr Zelin directed air operations personally by mobile phone from Moscow.*

In the run-up to the 9/11 terrorist attacks (Figure 7.4), there was no interoperability between the information technology and C2 systems of the Federal Aviation Administration (FAA) and the North American Aerospace Defense Command (NORAD).† In the immediate aftermath of the attacks, units of first responders on the ground often found that they were unable to communicate with each other. For example, the Port Authority Police Department had radios that could not talk to those of the FDNY.‡

In responses to major disasters and other emergencies, interoperability problems occur with depressing regularity. They are identified in carefully researched official after-action reports, only to occur again in future incidents.§ In the aftermath of Hurricane Katrina, the U.S. Department of Defense did not have an information-sharing protocol that might have enhanced communication and situational awareness between all the deployed military units. There were also major interoperability problems in communication between units of different federal, state, and local agencies on the ground. Joint Task Force Katrina, the National Guard, and the States of Louisiana and Mississippi did not have interoperable communications equipment.¶ Similar problems occurred in other disasters. During Australia's Black Saturday fires of 2009, the metropolitan and regional police forces had incompatible radio systems, and there was no interoperability between different

---

* McDermott (2009).
† Grant (2006).
‡ National Commission on Terrorist Attacks upon the United States (2004).
§ Donahue and Tuohy (2006).
¶ U.S. House of Representatives (2006); U.S. Senate (2006).

(b)

(a)

Figure 7.4 (a) A rescue worker near the World Trade Center destruction site from the top of one of the many trucks removing the collapsed rubble, September 14, 2001. (From the U.S. Federal Emergency Management Agency, http://www.fema.gov/media-library/assets/images/38874?id=3906. With permission.) (b) A U.S. Marine Corps helicopter hovers above the ground near a Soviet-built anti-aircraft weapon during the U.S. invasion of Grenada in October 1983. (From the U.S. Department of Defense, http://www.defenseimagery.mil, photo no. DF-ST-84-09905. With permission.)

emergency services agencies.* In the response to the 1987 King's Cross Underground fire in London, there was also no interoperability between the different emergency agencies, and between them and the London Underground. Despite being identified as a problem in the Fennell Report of 1988,† such difficulties recurred at least partially in the response to the 2005 "7/7" London bombings.‡

A study of 192 U.S. cities published in 2004,§ reported that 86% of them did not have interoperable communications with their state transportation department, 83% were not interoperable with the U.S. Department of Justice or the Department of Homeland Security, 60% were not interoperable with their state emergency operation centers, and 49% lacked interoperability with state police.

7.5.2.3 Security Constraints During military operations, there is an ever-present tension between the need to communicate information and the need for security and stealth. The proper balance is often very hard to reach. Sometimes, security procedures may restrict information availability to the point that the mission is harmed. This happened during the failed 1980 U.S. attempt to rescue the hostages being held at the U.S. Embassy in Iran, as discussed in Chapter 1. The security constraints further exacerbated a situation that was already plagued with structural organizational problems and communications interoperability issues.

German security procedures also played a role in delaying communications during the western offensive of 1914.¶

7.6 Summary of Factors

Table 7.3 summarizes the factors discussed above for the 20 incidents in Table 7.1. A glance at the table reveals that an *a priori* structural defect (organizational or systems) was a problem in almost all the cases, and there were communications shortfalls in most cases. Interoperability

---

* Parliament of Victoria, 2009 Victorian Bushfires Royal Commission (2010); Au (2011).
† Fennell (1988).
‡ *Guardian* (2011).
§ U.S. Conference of Mayors (2004).
¶ Fuhrmann et al. (2005).

Table 7.3 Situations Discussed: C2 Characterization

INCIDENT	MILITARY OPERATIONS / INABILITY TO COMMUNICATE								NOTES
	BECAUSE OF SYSTEM DESIGN OR POLICY SHORTFALLS					BECAUSE OF CIRCUMSTANCES			
	INAPPROPRIATE C2 APPROACH/ORGANIZATION DESIGN	BEHAVIORAL FAILURE TO COMMUNICATE	LACK OF INTEROPERABILITY	EQUIPMENT OR BANDWIDTH SHORTAGE	SECURITY CONSTRAINTS	INFRASTRUCTURE/ EQUIPMENT DESTRUCTION OR DAMAGE	PHYSICAL CONSTRAINTS	DENIAL BY ADVERSARY	
Great Retreat of 1914, First World War	■	■		■					[1]
German army in run-up to First Battle of the Marne, First World War		■		■	■			■	[2]
First Battle of Savo Island, Guadalcanal Campaign, Second World War	■	■							[3]
Mayaguez incident		■				■	■		[4]
U.S. hostage rescue mission	■	■	■	■			■		[5]
U.S. invasion of Grenada	■		■						[6]
First Gulf War, Operation Desert Storm, FSCL	■			■					[7]
Russia–Georgia War				■					[8]

(continued)

Table 7.3 Situations Discussed: C2 Characterization (*Continued*)

| INCIDENT | INABILITY TO COMMUNICATE | | | | | | | | NOTES |
| | BECAUSE OF SYSTEM DESIGN | | | | | BECAUSE OF CIRCUMSTANCES | | | |
	INAPPROPRIATE C2 APPROACH/ORGANIZATION DESIGN	BEHAVIORAL FAILURE TO COMMUNICATE	LACK OF INTEROPERABILITY	EQUIPMENT OR BANDWIDTH SHORTAGE	SECURITY CONSTRAINTS	INFRASTRUCTURE/ EQUIPMENT DESTRUCTION OR DAMAGE	PHYSICAL CONSTRAINTS	DENIAL BY ADVERSARY	
TERRORIST ATTACKS									
Oklahoma City bombing response	■			■					[9]
9/11 attacks response and possible prevention	■	■	■	■	■	■			[10]
7/7 London bombings response						■			[11]
2011 Norway attacks response		■							[12]
DISASTERS AND EMERGENCIES									
King's Cross underground fires response	■		■	■		■	■		[13]
Clapham Railway Junction accident response	■		■						[14]
Hillsborough stadium disaster response	■			■			■		[15]
Hurricane Andrew response	■		■	■		■	■		[16]
Columbine High School shootings response	■		■						[17]

								[18]
Indian Ocean Tsunami response								[18]
Hurricane Katrina response								[19]
Black Saturday fires response								[20]

Notes: Shaded box means heading is applicable to incident. [1] Gardner (2009); [2] Fuhrmann et al. (2005); [3] Hone (2006); [4] Toal (1998); Government Accountability Office (1976). [5] Anno and Einspahr (1988); Holloway (1980); Bowden (2006); Gass (1992); Thomas (1987); [6] Anno and Einspahr (1988); Cole (1997); [7] Tucker (2010); [8] Cohen and Hamilton (2011); McDermott (2009); Independent International Fact-Finding Mission on the Conflict in Georgia (2009). [9] Oklahoma Department of Civil Emergency Management (2003); [10] Grant (2006); National Commission on Terrorist Attacks upon the United States (2004); [11] Guardian (2011); Lieberman and Cheloukhine (2009); [12] Commission on 22 July (2012); Dennis (2012); [13] Fennell (1988); [14] Hidden (1989); [15] Hillsborough Independent Panel (2012); [16] Florida Governor's Disaster Planning and Response Review Committee (1992); [17] U.S. Fire Administration (1999); [18] Huber et al. (2008); [19] Moynihan (2006); U.S. House of Representatives (2006); U.S. Senate (2006); [20] Parliament of Victoria, 2009 Victorian Bushfires Royal Commission (2010); Au (2011).

problems were a specific factor less frequently, but were serious when they occurred. The inability to communicate because of security constraints was not particularly prevalent in the situations studied, but was disastrous in one case—the failed 1980 U.S. attempt to rescue the hostages in Iran.

Another pattern that emerges in Table 7.3 is that responses to major disasters such as Hurricane Katrina have been plagued across the board by all the problems identified here. The nearly total chaos produced by such disasters stretches all systems and organizations to the breaking point.

7.7 C2 Failure Model

In case studies such as the ones above, the lessons learned depend on the lens through which events are viewed. Some observers might look at the behaviors of specific individuals and note the decisions they made and how those decisions turned out. It is sometimes tempting to designate heroes and villains. Here, we look beyond individual deeds to see if we can find the underlying or root causes of the failure. For example, in the case where we find that critical information was not available in a timely manner to those who needed it, we look for the reasons why this occurred.

7.7.1 Causes of Failure

Understanding failure is a prerequisite for improvement. This requires looking beyond the particulars of the situation at hand to identify a risk that may be persistent and adversely impact future operations. Having a taxonomy of causes of failure contributes to one's ability to systematically analyze a failure and properly attribute causes.

We construct a "model of failure" one step at a time, beginning with Figure 7.5, which depicts a value chain for enterprise operations. If one assumes that an enterprise is capable of succeeding in a mission provided it takes appropriate and timely decisions and executes properly, then failures can be traced directly to damaged or broken links in this value chain. This is depicted by the arrows denoted as "Points of Failure." The Enterprise Approach depicted on the left of Figure 7.6 creates the conditions that shape enterprise capabilities

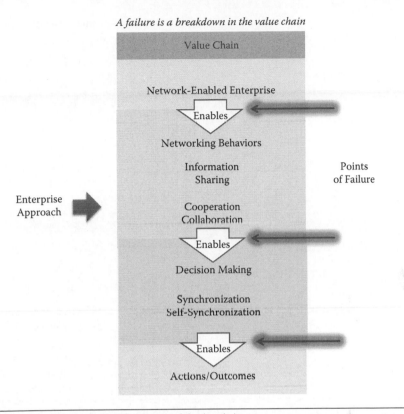

A failure is a breakdown in the value chain

Figure 7.5 The enterprise (C2) failure model value chain.

and behaviors, which in turn determine the strength of the links in the value chain and the degree to which they are vulnerable. Figure 7.6 depicts the relationships between the adopted Enterprise Approach and the sets of variables that constitute the value chain. This provides a way to trace the characteristics and performance of the enterprise.

Figure 7.7 categorizes C2-related failures by their relationship to the value chain. The first category, "Inability to Communicate," is associated with the characteristics and performance of the Enterprise Networks, as well as with the physical circumstances associated with the mission. The second category, "Failure to Share/Interact," is associated with the networking behaviors that are necessary to support decision making and subsequent action. The third category, "Decision Failures," is associated with the quality of decision making, the appropriateness of actions, and the ability to synchronize these actions.

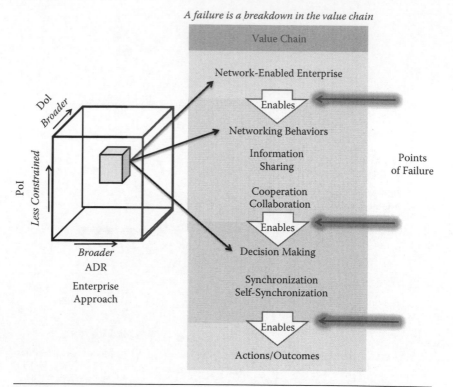

Figure 7.6 Enterprise (C2) failure model: impact of the Enterprise Approach. The Enterprise Approach determines the allocation of decision rights (ADR), patterns of interaction (PoI), and distribution of information (DoI).

7.7.2 Failures by Design

Ideally, the selection of an Enterprise Approach is the result of a conscious design process that considers the nature of the mission to be undertaken, the composition and capabilities of the enterprise, and the expected circumstances. Thus, the Enterprise Approach selected should be both feasible and appropriate for the situation, and not create impediments to success.

The design process is, at least in theory, supposed to take into consideration all the factors that could adversely impact success. If an enterprise functions as intended and nevertheless fails to accomplish the mission, this may represent a "failure by design": the Enterprise Approach may have been inappropriate to begin with.

Such inappropriate Enterprise Approaches are selected for a number of reasons, and it is important to understand these if we

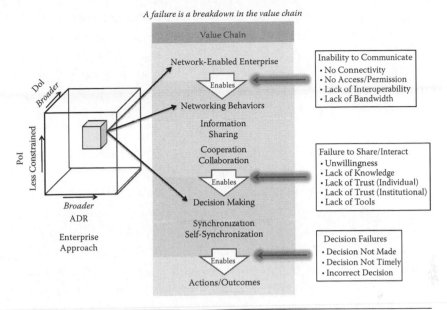

Figure 7.7   The enterprise (C2) failure model: C2 failure taxonomy.

are to develop effective enterprise remedies. The reasons include the following:

- The enterprise does not even recognize that the Enterprise Approach is a variable and adopts a "one-size-fits-all" approach.
- In a collective, entities focus only on their own organizations and not the enterprise as a whole.
- There is a misunderstanding of the nature of prevailing circumstances.
- The enterprise does not know which approach is appropriate for the mission and circumstances.
- The enterprise may know which approach might be best but is unable to adopt it.

Design problems often go undiagnosed when the spotlight is shined on the proximate cause of an operation's failure. As a result, attempts to fix the problem are often misdirected.

7.7.3 Failures Due to Changes in Circumstances

Even when an appropriate Enterprise Approach has been adopted and properly executed, a mission may fail. A major reason for this is that the mission or the circumstances or both may change. Such a change can result in the initially selected approach becoming inappropriate for the new situation or can prevent an enterprise from continuing to implement the approach that was selected. In either case, a different Enterprise Approach needs to be selected, and the organization needs to shift to the new approach in a timely manner. Enterprises that are able to (1) monitor circumstances, (2) recognize when changes can adversely impact mission success, and (3) make the necessary adjustments, including changing the Enterprise Approach in a timely manner, are displaying enterprise agility. As an enterprise becomes more agile, it can avoid some failures that would have otherwise occurred.

There are numerous ways that a mission or circumstances or both can change. Changes that are potentially significant are those that make an Enterprise Approach inappropriate or that interfere with an enterprise's ability to execute the approach successfully. The following list of potentially problematic changes is based on the work of NATO SAS-085[*]:

Changes to the Nature and Capabilities of the Enterprise (Self)

- Modification in the criteria by which an entity determines value
- Changes to acceptable bounds of performance (e.g., definition of mission success, constraints imposed on force employment or rules of engagement)
- Added or lost capability of self (e.g., deployment of a secure collaboration capability or the introduction of a disruptive technology)
- Degradation of system performance caused by physical damage, cyberattack, or some system failure that adversely affects performance (reduces one's ability to perform tasks)
- Modification in the composition of a coalition or collective (e.g., adding a new partner)
- Breach of information security (e.g., discovery of a Trojan horse)
- Loss of agreement or shared awareness among mission partners
- Loss of trust in information, information sources, or partners

---

[*] NATO (2013).

Changes to Mission/Environment (Operations)

- Different capability of an adversary (e.g., a new tactic, or a new disruptive technology or capability such as an offensive cyber capability)
- Different composition of the adversary (e.g., loss of an ally)
- Different operating conditions (e.g., terrain, weather)
- Changes in public perceptions of success or prospects of success
- New or emerging threat (e.g., change in government from friendly or neutral to adversarial)
- Change in mission scope and/or in the conditions on the ground (e.g., the loss of a permissive environment or the outbreak of a disease, which would make the problem more challenging)
- Change to the time available to accomplish a task (e.g., damage to an adversary that delays a planned attack)

7.7.4 Attribution of Cause

To determine if a particular failure was due to an initially faulty design (inappropriate approach) or due to a change in mission or circumstances that was not recognized and addressed by a corresponding change in the Enterprise Approach, or if a failure was not directly related to the approach employed, there is a need to look beyond proximate causes of failure (or instances where success is limited or comes at too high a price).

An observed lack of success could be the result of multiple causes, each of which could, even in isolation, be catastrophic. Such causes can be considered "independent," since the outcome is determined by any one of them without regard to the presence or absence of the others. These form the critical links in the value chain. Conditional causes are those that by themselves would not guarantee failure; however, they result in failure when they occur in some combination.

It is for this reason that Figure 7.5 through Figure 7.7 focus on the value chain. The Enterprise Approach (on the left) creates the conditions that shape individual and collective behaviors and place demands upon supporting systems. The behaviors shaped, facilitated, or constrained include not only individual, team, and organizational behaviors but also the behaviors of processes. These behaviors impact the performance of the systems that enable and support them. Collectively, these behaviors determine if the enterprise is functioning as intended. Many failures can be attributed to an enterprise that is not functioning as intended. However, failures also occur despite the fact that an enterprise is functioning as intended. Since there often are multiple causes of failure during actual operations, there can be instances of both of these types of failure, as we saw in many of the cases discussed above.

Failures can generally be traced to one or more damaged links in the value chain. The specific damages to the value chain are the proximate causes of the failure. Mission failures can be a result of the failure of a critical part of a system (a single point of failure), or they could be a result of a cascading set of failures (systemic risk). Since nothing works perfectly, and no enterprise is immune from surprise or misfortune, Enterprises routinely experience failures of various kinds. Success results when these failures can be successfully managed. It is not possible to prevent all mission failures. However, it is possible to reduce the likelihood of failure by eliminating or reducing failure by design. This means adopting the appropriate Enterprise Approach and ensuring the best technical performance of one's communications networks—paying particular attention to the interoperability of equipment.

7.7.5 Analysis of an Example in More Detail

One of the cases discussed earlier in this chapter is the German western offensive of 1914. Our analysis here goes beyond simply attributing blame to the three senior commanders, in an effort to illuminate the possibility that the C2 Approach employed was inappropriate given the situation and contributed to the conditions for failure.

A major failure in the German western offensive of 1914 involved two large field armies attempting to execute a modified version of the Schlieffen Plan. The plan was designed to attack France quickly through neutral Belgium before turning southward to encircle the French army on the German border. For their part, the French employed "Plan XVII" with the objective of recapturing Alsace-Lorraine. The Germans adopted an approach to C2 that they had

Figure 7.8   General von Kluck's army marching in France, 1914. (From http://pierreswesternfront. punt.nl/content/2008/08/marne-verberie-nery-villers-cotterets. With permission.)

successfully utilized previously in the Franco-Prussian War, featuring at least a partial application of mission-type orders (*Auftragstaktik)* and some presumption of shared intent and awareness (see Chapter 6). However, the situation faced in World War I differed significantly from that of the Franco-Prussian War. First, the size of the battlefield was much greater in World War I, with German armies (Figure 7.8) spread over hundreds of miles. The Germans recognized this difference and employed wireless communications (then a new and advanced technology) along with messengers in motor vehicles (another relatively advanced technology for the time) in an attempt to maintain shared awareness. As it turned out, clogged roads and a French wireless jamming station on top of the Eiffel Tower, along with German security procedures for handling messages, denied the Germans the rapid communications they needed to maintain shared awareness.

If the offensive had gone according to plan, this lack of real-time shared awareness might, possibly, not have become a fatal problem. We should note that there is considerable disagreement over whether or not the Schlieffen Plan, or its modified version, was ever workable at all, even under ideal conditions. In any case, there were major deviations from the plan, and more importantly, those deviations were uncoordinated. Although General Von Kluck[*] made his turn to the Southeast partly to support Field Marshall von Bülow,[†] suggesting at

---

[*] Alexander Heinrich Rudolph von Kluck (1846–1934).
[†] Karl von Bülow (1846–1921).

least some amount of coordination between commanders, subsequent actions were poorly coordinated, especially with headquarters. For example, headquarters ordered a general retreat for the two armies based on its perception of von Bülow's condition, unaware that von Kluck had improved his own position. Meanwhile, Prince Rupprecht of Bavaria's* counterattack in Lorraine, while known to headquarters, was not well coordinated with the operations of the other German armies. As a result of all this, the French forces were concentrated rather than split and the offensive was ultimately unsuccessful.†

There are of course many ways to assign "blame." Were the decisions of General Von Kluck and Field Marshall von Bülow to deviate from the plan justified by circumstances? Perhaps they may have been, if we consider only the local circumstances of their forces. Having decided to deviate, could they have coordinated their subsequent actions better with headquarters and each other? Should headquarters have restrained Prince Rupprecht when he made his demand to counterattack in Lorraine? Could the counterattack have been coordinated with other parts of the German forces? Should a "Plan B" in case of communications failures have been developed? Should headquarters have recognized the consequences of deviations to the plan and taken appropriate action?

Rather than stop with a finding of fault with one or more of the decisions made, a more comprehensive approach to agility analysis is warranted. This is because mistakes will always be made, circumstances will not be as expected, and circumstances will change. One of the most quoted remarks along these lines comes from one of the fathers of mission command, von Moltke the Elder: "No plan of operations extends with certainty beyond the first encounter with the enemy's main strength," more often shortened to "No plan survives contact with the enemy."‡ Given this, is it reasonable to ask a series of questions that allow us to ascertain the degree to which the German headquarters and field commanders were collectively agile in the face of change? Collectively, agility was required because the situation was complex and thus could not be decomposed into independent parts. In fact, individual agility in the absence of shared awareness was a

---

* Rupprecht or Rupert, Crown Prince of Bavaria (1869–1955).
† Fuhrmann et al. (2005).
‡ Hughes (1993).

Approach Agility Questions?	German WWI Western Offensive, 1914
Initial Enterprise Approach appropriate?	Possibly: Modified Schlieffen Plan with Auftragstaktik, and shared awareness ensured by virtue of communications capabilities
Did mission or circumstances change?	Yes, in two significant ways: Uncoordinated deviations from plan and denied communications
Were these changes significant?	Yes, The situation was complex in that the tasks assigned to the individual field commanders could and were impacted by each other and thus were interdependent.
Was Enterprise Approach still appropriate?	No A lack of communications resilience meant that shared awareness could not be regained. Thus the allocation of decision rights needed to be changed and a new plan developed and implemented
Was there self-monitoring?	No They did not consider approach a control variable
Was the need for a new approach recognized?	No
Was a more appropriate approach identified/available?	N/A
Was enterprise able to adopt a more appropriate approach in a timely manner?	No

Figure 7.9 Mission analysis: unsuccessful mission.

cause of the problem. Figure 7.9 summarizes the results of a failure analysis from an Enterprise Approach perspective.

While the Enterprise Approach selected for the offensive may have been appropriate, its appropriateness was dependent upon three assumptions. The first assumption was that the exercise of the decision rights allocated to field commanders would not result in decisions that adversely affected one another or the operation as a whole. The second assumption was that adequate communications capability would ensure the development of shared awareness that was timely enough for the circumstances. The third was that any changes in mission and circumstance would not require a change in the Enterprise Approach. As it turned out, none of these assumptions proved to be true.

Even if adequate communications had been maintained, or restored after a disruption, it is not clear whether the decisions of field commanders or headquarters would have been different. It is not clear that the actors fully recognized the complexity of the situation, and that the resulting interdependencies required the timely development of shared awareness. It seems, with hindsight, that the Germans under-estimated the need for coordination that their plans generated. Simply stated, they were not aware of the vulnerabilities of the Enterprise Approach adopted, the need to monitor it to make sure it was still appropriate, and of options to modify their approach. Thus, the failure of the German offensive could be attributed, at least in part, to a lack of Enterprise Approach agility.

This diagnosis of a lack of Enterprise Approach agility leads to a different and more comprehensive set of remedies than if the failure was attributed to leadership mistakes or to a lack of communications capability. For example, if one is convinced that the failure experienced was caused by an inability to communicate, then one might see investments in better equipment as a sufficient remedy. If one believes that the problem also has a behavioral component (rather than simply being a technical problem), then one might look to changes in education, training, and doctrine.

7.7.6 *Applying the Analysis to an Example of Success*

In Chapter 1, we discussed Nelson's victory at Trafalgar. Figure 7.10 shows the application of the type of analysis illustrated above to that case.

Approach Agility Questions?	Nelson at Trafalgar
Initial Enterprise Approach appropriate?	Yes: Assuming traditional naval tactics
Did mission or circumstances change?	Yes: A new tactic was needed to achieve mission success
Were these changes significant?	Yes
Was the Enterprise Approach still appropriate?	No: The new tactic was not compatible with the traditional Enterprise Approach because it interfered with the lines of sight needed for communications
Was there self-monitoring?	Yes: In the form of conscious co-evolution
Was the need for a new approach recognized?	Yes
Was a more appropriate approach identified/available?	Yes
Was the enterprise able to adopt a more appropriate approach in a timely manner?	Yes

Figure 7.10 Mission analysis: successful mission.

7.8 Final Remarks

The above analysis suggests that in many cases, C2 failure may be avoided by adopting the appropriate Enterprise Approach and adjusting that approach in an agile manner as circumstances change. In Chapter 8, we further explore the concept of C2 Agility, and which approaches might better lend themselves to it, by examining experimental evidence.

The case studies above also underscore the extreme importance of effective communications for C2. Chapter 8 discusses the resilience of various C2 Approaches to degradations in communications and information capability, again in the light of experimental evidence.

EXPERIMENTAL EVIDENCE
AND ANALYSIS

So far, we have described and analyzed the four megatrends shaping the future of Command and Control (C2), and indeed enterprise problem solving in general. We have also analyzed many of the ways that C2 can go wrong, some of which are systemic.

Our most fundamental argument is that to succeed, an enterprise must be agile, especially in the context of Big Problems, and that agility is enabled by a Robustly Networked Environment, advanced information and communications technology (ICT), and the availability of several alternative organizational forms.

This means the following:

- Having a range of C2/management/governance approach options available, particularly ones that are network enabled
- Knowing which option is appropriate for the circumstances at hand
- Being able to transition from one Enterprise Approach to another in a timely and efficient manner
- Preventing a desired Enterprise Approach from being denied

8.1 Hypotheses

In various sections of this book, we have presented a number of assertions about the nature of organizational design, and the properties and merits of different Enterprise Approaches. In this section we treat these as testable hypotheses, present some of the experimental evidence we have found, and present our conclusions.

It is important to emphasize that the experiments described in the sections below do not necessarily represent "ground truth." Experiments are obviously not the same as real military or disaster relief operations, with all the attendant risks and real-world consequences. On the other hand, examining real operations, as we have done elsewhere in this book, also has limitations, because cases vary so widely in conditions and circumstances, and there are so many variables whose effects are mixed together. What the experiments offer us is the opportunity to isolate some variables and study their effects, hence yielding insight. They offer additional suggestions about whether some of our hypotheses about organizations and C2 Approaches, gleaned from various case studies and sometimes common sense, have merit.

The hypotheses fall into the following groups:

- Hypotheses about organizational designs and Enterprise/C2 Approaches:
  - H1: There are many approaches to command and control (military) and organizational design (military and civilian).
  - H2: Different C2 Approaches and organizational designs map to different regions of the Enterprise Approach Space.
  - H3: Traditional hierarchies are located near one corner of the Approach Space, and fully connected edge organizations are located near the opposite corner.
  - H4: There is no universal C2 Approach or organizational design that is appropriate for all problems (tasks) and circumstances.

- Hypotheses about Traditional Hierarchies:
  - H5: Traditional hierarchical organizations are well adapted to some problems and circumstances but are not as well adapted as other organizational approaches for other problems and circumstances.
  - H6: Traditional hierarchies are often, but not necessarily always, the most appropriate Enterprise Approach for Industrial Age problems.
  - H7: Traditional hierarchies do not typically perform well for problems that are complex and dynamic (Big Problems).

- Hypotheses about Decentralized, Net-Enabled ("Edge") Organizations:
  - H8: Edge organizations are well adapted to some problems and circumstances but are not as well adapted as other Enterprise Approaches for other problems and circumstances.
  - H9: Edge approaches are often the most appropriate choice for Big Problems.
  - H10: More network-enabled approaches are more appropriate for Big Problems than less network-enabled approaches.

- Hypotheses about Properties of C2 Approaches:
  - H11: Approaches that are more network enabled also tend to be more agile.
  - H12: Balanced approaches are more effective than unbalanced ones.

- Hypotheses about organizational design choices and how they can be constrained by capabilities and circumstances:
  - H13: The quality of the available information limits the potential effectiveness and efficiency of an enterprise.
  - H14: Some Enterprise Approaches are better able to take advantage of high-quality information, or perform despite low-quality information, than others.
  - H15: Communications and information system capabilities can limit which approaches are feasible or appropriate.
  - H16: Trust between and among individuals or organizational components can limit approach options.
  - H17: Attacks that degrade communications and information capabilities impact network-enabled approaches more than traditional hierarchies.

- Hypothesis about Enterprise C2 Agility:
  - H18: Changes in circumstances can require a change in Enterprise Approach, and an agile enterprise must be able to execute such a change.

8.2 Nature of Experimental Evidence

Both case studies (retrospective analysis of actual events) and experiments (instrumented, controllable environments) can provide evidence that can support or refute hypotheses such as the ones enumerated in Section 8.1. Evidence from both case studies and experiments was analyzed by an international team of researchers that was chartered under the auspices of NATO's Science and Technology Organization (STO)[*] to validate a conceptual model (theory) of C2 Agility. These findings, conducted by the NATO research group SAS-085,[†] provide some evidence supporting the various assertions we make in Section 8.1. One of the authors of this book (David Alberts) served as chairman of NATO SAS-085 and was a principal author of the group's final report.[‡] The material in this section, about the nature of the NATO experiments and their meta-analysis,[§] is *taken verbatim or closely paraphrased from that report*, to which the reader desiring further detail is referred.

The SAS-085 experiments were all agent-based model simulations. Each simulation-based experiment exploited an experimental platform (a constructive simulation tool configured in a specific way within the context of an operational or mission setting) capable of instantiating two or more of the C2 Approaches described in the NATO Network Enabled Capability (NEC) C2 Maturity Model (often abbreviated to N2C2M2)[¶] under a set of circumstances. In addition, the diversity of missions and scenario contexts offered by all the experiments presented a unique opportunity to test the same set of hypotheses in a broader context. Therefore, in order to produce a more complete, robust, and generalizable set of findings,

---

[*] http://www.sto.nato.int/.

[†] NATO System and Analysis Studies (SAS) 085: C2 Agility; https://www.cso.nato.int/ACTIVITY_META.asp?ACT=1912.

[‡] NATO (2013).

[§] Mega-analysis, pooled analysis, and meta-analysis (Bravata and Olkin, 2001) are three methods for combining data and/or results from various experiments (Curran and Hussong, 2009). The ability to increase the sample size and the variety of data has many advantages, including increased statistical power, reduced exposure to local biases, and in the case of meta-analysis, improved control of between-study variations. SAS-085 Experimentation Team used meta-analysis for the C2 Agility Campaign of Experiments.

[¶] Alberts et al. (2010).

rather than looking at the results of each single experiment in isolation, the SAS-085 Experimentation Team formulated a Campaign of Experimentation that employed a prospective meta-analysis across the multiple simulation-based experiments.

This approach allowed SAS-085 to (1) control or influence the values of key variables of interest (e.g., the C2 Approach); (2) explore a range of missions and circumstances; and (3) collect a great deal of detailed information about behavior and consequences. An advantage of employing multiple simulation models and experimental venues is that this facilitates the exploration of a large and diverse Endeavor Space (discussed later in this chapter) and provides a better estimate of the potential agility of various C2 Approaches. Furthermore, it strengthens the tests of statistical significance and makes it possible to generalize the findings beyond that which would be appropriate if only one model were employed. The series of experiments conducted by SAS-085 looked at hundreds of mission-circumstance pairings and collected detailed information about behaviors and results. Each run was mapped to a position in the C2 Approach Space, thus providing an opportunity in the meta-analysis to look at hypotheses that the case studies could not. In addition, the experimental results were used to calculate the potential agility of each of the C2 Approaches and, assuming perfect C2 maneuver agility, the agility of different levels of C2 Maturity.

SAS-085 selected a total of six constructive simulations[*] to employ in the C2 Agility Campaign of Experimentation. These six had been previously used to conduct at least one experiment whose objectives were compatible with those of SAS-085's experimental campaign. The simulations that provided data for the meta-analysis were ELICIT-IDA (USA), ELICIT-Trust (USA), abELICIT (Portugal), IMAGE (Canada), WISE (UK), and PANOPEA (Italy). The following paragraphs briefly describe each of these experiments.

---

[*] *Constructive simulation* is defined as modeling and simulation involving simulated people operating simulated systems.

8.2.1 ELICIT Experiments

The Experimental Laboratory for the Investigation of Collaboration, Information-Sharing and Trust (ELICIT) platform* was originally developed by the U.S. Department of Defense's (DoD) Command and Control Research Program† to facilitate the testing of hypotheses related to edge and hierarchical (traditional) approaches to C2. ELICIT is a virtual instrumented environment where humans are connected over a configurable network to accomplish assigned tasks. ELICIT can be used to explore empirically the relationship among approaches to C2 and organization, team and individual characteristics, and value-related measures in the network-centric value chain. ELICIT has been enhanced with the addition of software agents that can "stand in" for human participants.

When all of the participants are software agents, ELICIT can run faster than real time. Three applications of agent-based ELICIT have been used in support of SAS-085's Campaign of Experimentation. Two of these, ELICIT-IDA‡ and abELICIT,§ were used to explore the relative agility of four C2 Approaches. Success for a simulated shared awareness task was defined as satisfying a combination of correctness and timeliness requirements (how many individuals correctly solved the problem, and the time they required). Efficiency was also measured. The C2 Approaches were evaluated using a variety of mission challenges and stresses, which are different for ELICIT-IDA (see below) and abELICIT.

The third application of agent-based ELICIT, ELICIT-Trust, is a variant in which the agents evaluate the behavior of other nodes and generate an estimate of trust for those agents with which they interact.¶ Trust is evaluated based on the willingness of another agent to provide information and the competence of the agent to provide valuable information. Based on these estimates of trust, agents adapt their behaviors as they seek to interact with the most trustworthy agents, with the goal of maximizing mission performance.

---

* Ruddy (2007).
† http://www.dodccrp.org; CCRP (2010).
‡ Alberts (2011).
§ Alberts and Manso (2012).
¶ Chan and Adali (2012); Chan et al. (2012).

All ELICIT applications for SAS-085 used a scenario challenging human participants or software agents to find the "Who, What, Where, and When" of a terrorist attack. Throughout the course of the experiment, "factoids" (i.e., information elements that are pieces of the puzzle) are distributed to participants or agents. If permitted to do so by the configuration of their organization and approach to C2, participants or agents may (or may not) disseminate factoids to other participants or agents by sharing information and collaborating using this instrumented platform. However, only by sharing information can they achieve sufficient levels of awareness to solve the problem completely. The challenge in ELICIT is unambiguous and clear: it has a clearly defined objective and all information is accessible. Moreover, the factoid set (i.e., the set containing all factoids) has no ambiguity among the factoids, and it does not contain erroneous information. Nonetheless, the dynamics caused by human participants or agents during the runs often result in complex behaviors. For example, the order in which factoids are received by agents (in turn, a function of individual agent decisions) affects the final results.

The three ELICIT applications instantiated C2 Approaches (see Chapter 6 and Section 8.3) following the organizational options depicted in Figure 8.1, extracted from Alberts (2011). However, each experiment implemented slightly different variations of these organizations by, for example, changing the number of team members and leaders. Experiments also differed in how they defined acceptable performance or "success." For example, ELICIT-IDA maintained the same measures of effectiveness (number of individuals who correctly solved the problem and the time it took for the first correct solution to be developed) for all C2 Approaches and used mission requirements as the standard for determining success, while abELICIT evaluated success differently for each C2 Approach. In runs where the entity employed either a conflicted or a de-conflicted C2 Approach, success depended on the ability of all team leaders to solve the problem. When a coordinated C2 Approach was employed, organization success depended on the coordinator finding the correct solution. When a collaborative C2 Approach was employed, organization success depended on the coordinator finding the correct solution to all problem spaces or team leaders finding the correct solution to their own problem spaces.

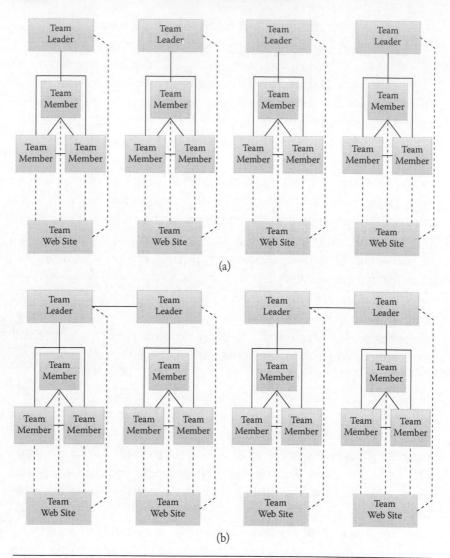

Figure 8.1   ELICIT instantiation of C2 Approaches: (a) conflicted, (b) de-conflicted, (c) coordinated, (d) collaborative, and (e) edge. (From Alberts, 2011. With permission.) (*continued*)

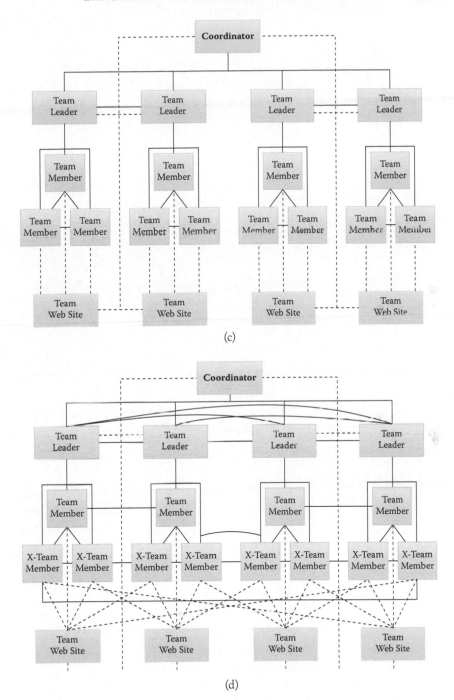

Figure 8.1   (*continued*) ELICIT instantiation of C2 Approaches: (a) conflicted, (b) de-conflicted, (c) coordinated, (d) collaborative, and (e) edge. (From Alberts, 2011. With permission.) (*continued*)

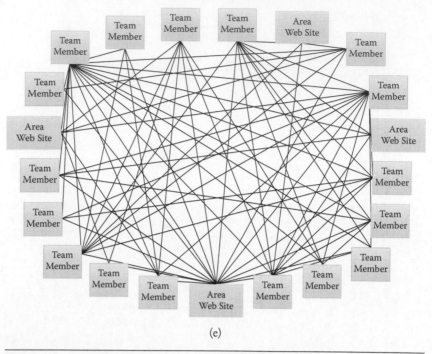

(e)

Figure 8.1   (*continued*) ELICIT instantiation of C2 Approaches: (a) conflicted, (b) de-conflicted, (c) coordinated, (d) collaborative, and (e) edge. (From Alberts, 2011. With permission.)

Finally, when an edge C2 Approach was employed, organization success depended on a plurality of individuals being correct.

8.2.2 IMAGE Experiments

IMAGE* was developed as a suite of generic representations, and "scenarization," simulation, and visualization tools aimed at improving the understanding of complex situations. More recently, a simulation-based experiment was designed with IMAGE to investigate how collective C2 Approaches (see Chapter 6 and Section 8.3) instantiated in a specific operational context impact agility and mission effectiveness.† IMAGE represents organizations that are implemented by software agents that deliberate and act according to rules that comply as much as possible with the collective C2 Approaches described in Chapter 6 and Section 8.3.

---

* Lizotte et al. (2008); Lizotte et al. (2013).
† Bernier (2012).

The scenario chosen for SAS-085 simulates a failing state that has experienced years of civil war and conflicts with a neighboring country. The central government and local authorities have been struggling with rebels, refugees, poverty, and starvation for many years. The simulation begins with the arrival of the international community in the form of military establishments, other government departments (OGDs), and nongovernmental organizations (NGOs). Their mandate is to secure and stabilize the failing state. Each organization within the scenario conducts activities according to their area of responsibility. For example, a military organization is responsible for providing security for economic development or humanitarian activities being conducted by NGOs. Reflecting reality, the scenario is designed such that cooperation between organizations significantly improves the likelihood of success. The consequences of cooperation are less conflicting actions and higher levels of synergy. Such an approach is called "comprehensive."[*] The IMAGE experimental platform supports a variety of challenges and circumstances within the scenario, including information-sharing delays, organizations that retract unexpectedly, enemy strengthening, and surges in crisis severity. The combination of all these changes produces a scenario that represents an Endeavor Space comprising 54 circumstances. Each C2 Approach is exposed to the whole Endeavor Space, thereby revealing comparative advantages and weaknesses in terms of C2 Agility, which in turn supports the assessment of the hypotheses.

IMAGE instantiated each of the C2 Approaches by configuring the Allocation of Decision Rights, Patterns of Interaction, and Distribution of Information (ADR, PoI, and DoI) between organizations. Figure 8.2 provides a description of how the collective implemented each C2 Approach. More details are provided in Bernier (2012).

8.2.3 WISE Experiments

The Wargame Infrastructure and Simulation Environment (WISE)[†] is a land-focused C2 model with representation of air and maritime support to land operations at the system level. WISE represents

---

[*] Leslie et al. (2008).
[†] Pearce et al. (2003).

C2 APPROACH	ADR	POI	DOI	PLANNING PROCESS
Conflicted	Each organization decides on its unit locations and activities	Between units of the same organization	Between units of the same organization	Move units(s) to most problematic province(s) and then select the activity for each unmoved unit that impacts the variable with the lowest value
De-Conflicted	Each organization decides on its unit locations and non-conflicting activities	With organizations having collocated units for preventing conflicting activities	Variables shared instantly between organizations having collocated units	Like in *Conflicted* but conflicting activities are not allowed
Coordinated	Like in D*e-Conflicted* but interacting activities are considered first with collocated units	With organizations having collocated units for considering interacting activities	Like in D*e-Conflicted* + variables shared with 5 non-collocated units (delay: 5 iter.)	Like in *Conflicted* but all possible interactions between activities with collocated units are considered
Collaborative	All activities and unit locations are decided collectively	With all organizations for deciding unit locations and activities	Same as *Coordinated* but with any number of units (delay: 3 iter.)	All combinations of unit locations and activities are considered; those with the higher impact are retained

Figure 8.2   IMAGE instantiation of C2 Approaches. (From NATO, 2013.)

warfighting and—currently in a limited capacity—peace support or stabilization operations, operating either as a human-in-the-loop war game or as a closed-form constructive simulation. The basic conceptual framework within WISE is built around organizations that are used to represent either individual entities, such as a single tank, or aggregated units, such as a company. A scenario typically instantiates a mixture of the two with the war game being used to provide the basis for follow-on constructive simulation runs. Within WISE, organizations are sufficiently generic in nature to allow for a wide range of scenario actors to be represented. For example, an organization can represent an individual platoon, unmanned air vehicle, improvised explosive device, insurgent force, groups of civilians, and so on. WISE is a C2-centric model in that it has a detailed representation of communications and networks and the information flows across those networks, leading to the development of a perception against which an organization makes decisions. These perceptions are measured to determine the level of situation awareness present at an organization, thereby enabling studies of C2 by examining the impact of changes in C2 capability through the measures of merit hierarchy.* The C2-centric nature of WISE, combined with the representation of communications and networks, enables a number of changes of circumstances, such as perturbations or enhancements to C2, to be represented.

The scenario chosen for SAS-085 is similar to that chosen for IMAGE with WISE simulating a failing state that is experiencing internal conflict. The central government has invited a NATO Coalition to stabilize the country. The UK operation represents a brigade-size operation with the specific intent of clearing insurgents from a major urban area. This task falls to a single battle group with other battle groups performing security and isolation tasks. The simulation represents a range of complex factors, including civilians, insurgent attacks across the area of operations, and a range of other destabilizing actions ensuring a dynamic and complex scenario. Success for the UK brigade is the defeat of insurgents within the urban area.

Two C2 Approaches were represented within WISE: de-conflicted and collaborative. The representation of de-conflicted C2 within WISE represented patterns of interaction and the distribution of information organized along boundaries and areas of responsibility (i.e., each battle group was assigned its own Area of Operation [AO]). The C2 links between subunits and battle group headquarters (HQ), and between battle group HQs and brigade HQ, were hierarchical with no peer-to-peer links to other battle group HQs. ADR was represented through joint fires assets being controlled at both battle group HQ and brigade HQ with mortars at company HQ; there was no sharing of pooled resources. Finally, the rules of engagement were

---

* Fellows et al. (2010).

tightened to represent reduced availability of information and hence more uncertainty in the targets being selected for engagement. The representation of collaborative C2 within WISE represented patterns of interaction and the distribution of information up the command hierarchy as in the de-conflicted approach but also peer-to-peer links across company HQs and battle group HQs. ADR was represented through joint fires assets being shared and resources pooled at brigade HQ to enable targets to be prioritized across the different battle group areas of operation. Mortars, however, were still held at company HQ. Finally, the rules of engagement were relaxed to represent greater availability of information and hence more certainty in the targets being selected for engagement.

Figure 8.3 depicts the way the different C2 Approaches were implemented in WISE.

8.2.4 PANOPEA Experiments

The Piracy Asymmetric Naval Operation Patterns modeling for Education and Analysis (PANOPEA)* is an agent-based simulation that directs IA-CGF (Intelligent Agents Computer Generated Forces) through the application of a range of strategies and based on their situation awareness to successfully defeat pirates. PANOPEA allows different C2 Approaches to be instantiated. For example, in this series of experiments, de-conflicted, collaborative, and edge C2 Approaches were instantiated.

The scenario chosen for SAS-085 is one involving piracy in the Horn of Africa. The scenario includes naval vessels and helicopters, intelligence assets, ground bases, cargo ships, other boats (i.e., fishing boats and yachts), as well as pirates hiding in the general traffic. The simulation represents a range of complex issues, including pirate attacks on navy and cargo vessels, information flow, and actions by friendly forces to deter or defeat pirate attacks. In a way that is similar to the other experimental platforms, PANOPEA supports a wide variety of challenges and circumstances, including variations in decision-making capabilities, weather conditions, the effect of misleading information, and the number of pirates. Altogether, these create a

---

* Bruzzone et al. (2011a,b).

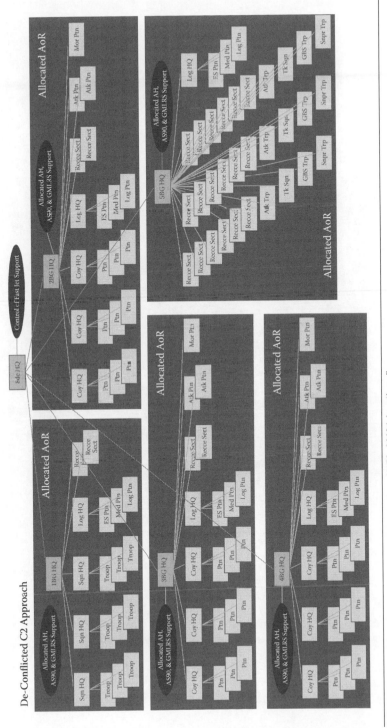

Figure 8.3 WISE instantiation of C2 Approaches. (From NATO, 2013.) *(continued)*

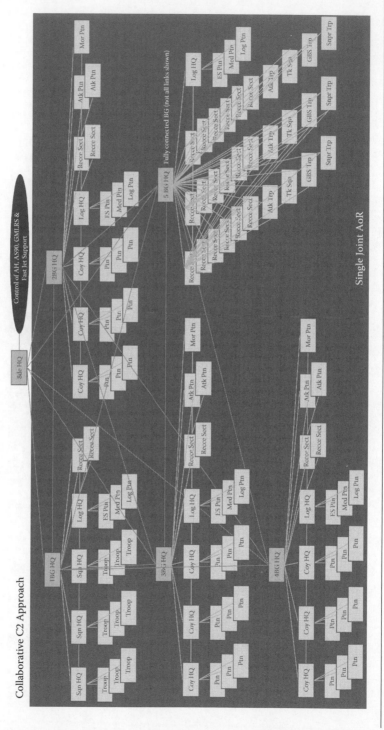

Figure 8.3 *(continued)* WISE instantiation of C2 Approaches. (From NATO, 2013.)

complex Endeavor Space against which each C2 Approach can be tested.

PANOPEA instantiated three C2 Approaches: de-conflicted, collaborative, and edge. The PANOPEA scenario for SAS-085 experimentation was designed with two coalitions operating in the area. Frigates are controlled through a command chain. Information is provided by ship intelligence, cargo vessels, and local authorities. Actors having different roles interact and take decisions according to the predetermined configurations illustrated in Figure 8.4. The level of connectivity increases when moving from de-conflicted to collaborative and from collaborative to edge.

8.3 Empirical Support for Hypotheses about Organizational Designs and Enterprise/C2 Approaches

In the sections that follow, evidence and/or logical arguments are presented to support the hypotheses presented above.

H1: There are many approaches to command and control (military) and organizational design (military and civilian).

Command is the expression of the challenge at hand. It is essentially a formulation of the problem to include an articulation of the objective function, the identification of the controllable variables, and the establishment of constraints. Control is exercised by the establishment of structures and processes that are enabled by systems. Military organizations have successfully employed different approaches to C2 in the past and continue to experiment with tailoring their approaches to the situation at hand.

An absolutely critical research finding from the Department of Defense's Command and Control Research Program[*] has been that, during the Industrial Age, there has not been a single best approach to, or philosophy of, C2.[†,‡] Had a best approach been found, one could argue that there should be only one approach employed and that other approaches, while possible, are not options truly worthy of consideration.

---

[*] http://www.dodccrp.org.
[†] Alberts and Hayes (2003), pp. 18–19.
[‡] Also see Alberts and Hayes (1995), pp. 77–100; Alberts and Hayes (2006), pp. 169–180.

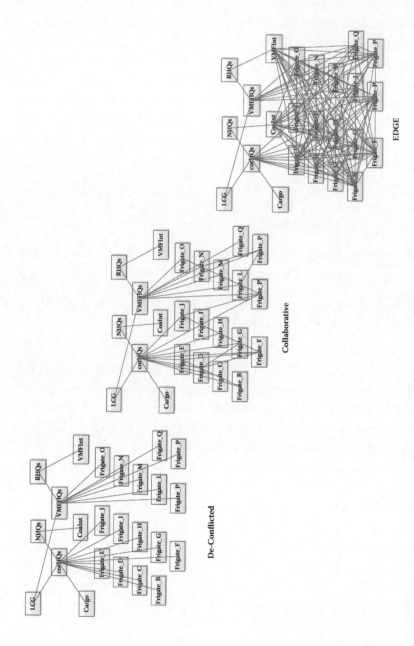

Figure 8.4 PANOPEA instantiation of C2 Approaches. (From NATO, 2013.)

One key dimension of C2 and organizational design is the way decision rights are allocated—that is, who gets to make what decisions. In Chapter 6, we discussed mission command, a philosophy of command that addresses the delegation of decision rights and has a rich history dating back to the 19th century. In earlier times, a major reason for delegation was the absence of advanced communications technologies (e.g., Nelson at Trafalgar, or the various wars fought by the Prussians in the 19th century). Today, there is a presumption of connectivity, and a temptation for higher echelons far removed from the action to micromanage operations. This temptation can occur for a number of reasons, including political and legal constraints. The availability of communications for situational awareness, and the avoidance of destructive micromanagement, are now prime motivators for emphasizing mission command. The other major reason for stressing mission command is that appropriate delegation allows organizations to operate more effectively in complex and dynamic situations that cannot entirely be anticipated and prepared for in detail. Mission command does not represent a single C2 Approach. Rather, it denotes an overall philosophy of command, with varying breadths of Allocation of Decision Rights depending on circumstances. Thus, it encompasses a rather large family of C2 Approaches. As we discussed in Chapter 6, the Allocation of Decision Rights in mission command can also drive changes in the Patterns of Interaction and the Distribution of Information.

Understanding different C2 Approaches and knowing their strengths and weaknesses is key to being able to operationalize a philosophy of mission command. Until recently, there was no conceptual framework to facilitate our understanding of the nuances of different approaches to C2 and to systematically design and compare approaches to one another. C2 Approaches were described simplistically as being located somewhere along a centralized–decentralized continuum. A commonly used expression, "centralized command–decentralized execution," is sometimes thought to represent a specific approach to C2. In reality, this refers to some delegation of authorities, but it is rather vague, amounting to nothing more than a type of mission command by another name. In this book, we often use the terms "decentralized" and "net-enabled" interchangeably, but this is because we view decentralization as occurring across more dimensions than just the allocation of decision rights.

Recalling the discussion of Chapter 6, C2 Approaches differ from one another along the following three dimensions:

- Allocation of decision rights
- Patterns of interaction
- Distribution of information

A particular approach to C2 corresponds to a region of the C2 Approach Space depicted in Figure 8.5, repeated from Chapter 6.

Policies, practices, processes, and commander decisions directly or indirectly impact each of these dimensions. Furthermore, these dimensions are interdependent, with behaviors being enabled or constrained as a function of the policies, practices, and processes of an organization and command decisions. For example, the way decision rights are allocated can have a considerable influence on information flows and interaction patterns. The availability of information will impact patterns of interaction. The ability of an approach to C2 to contribute to the organization's mission depends upon achieving an appropriate

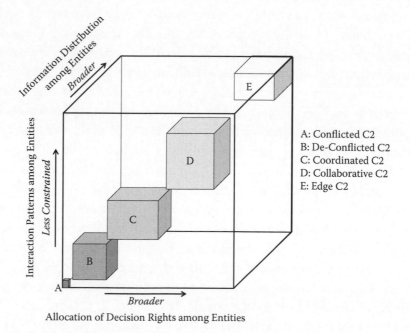

Figure 8.5 Repeated from Chapter 6. C2 Approach Space for collective endeavors. Positions are illustrative and not intended to be taken as precise.

balance between and among these dimensions. "Decentralization" in C2 involves all of them.

H2: Different C2 Approaches and organizational designs map to different regions of the Enterprise Approach Space.

The advances in ICT, as discussed in Chapter 4, have enabled widespread sharing of information within and across organizations. The United States and its NATO allies recognized that these advances afforded them the opportunities to rethink how the functions associated with C2 could be accomplished, and committed themselves to a "network-enabled capability." They recognized that they could not accomplish this information-age transformation all at once and needed to approach it incrementally.

NATO Research Group SAS-065* was formed to define a step-by-step "maturity model" for network-enabled C2 for collectives of multiple-partner organizations. Building upon previous work they employed the three-dimensional Approach Space we discussed in Chapter 6, and graphically located within this space a set of different C2 Approaches that spanned the range from less to more network enabled—that is—from centralized hierarchies and coordinated approaches to increasingly distributed and decentralized collaborative and edge approaches. Figure 8.5, repeated from Chapter 6, shows that these approaches do not overlap, and that each approach occupies its own "territory" along a diagonal of the C2 Approach Space. SAS-065 provided only general descriptions of the characteristics of each of these approaches. The boundaries between adjacent approaches are not precise.

A follow-on NATO research group, SAS-085, discussed in Section 8.2, developed quantitative scales for each of the three dimensions of the C2 Approach Space, and based upon the SAS-065 verbal descriptions of each C2 Approach, instantiated at least two of these approaches in different simulation models. Simulation runs were made under a variety of conditions. The data collected enabled the researchers to obtain a single point in the three-dimensional space for each

---

* NATO System and Analysis Studies (SAS) 065: Network Enabled Capability (NNEC) C2 Maturity Model; http://www.cso.nato.int/Activity_Meta. asp?Act=1382.

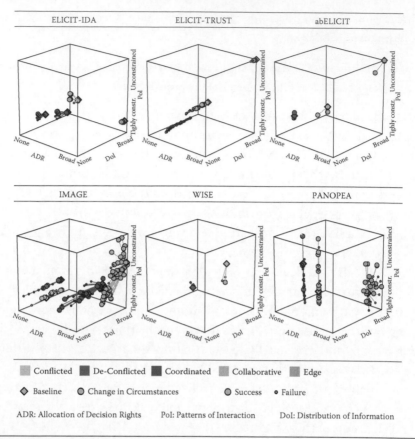

Figure 8.6 Positions in the C2 Approach Space. (From Alberts et al., 2013.)

approach under each condition for each model. These points allow us to visualize the shapes and locations of the regions occupied by each of the approaches as they were instantiated in each of the models, and compare them to the "idealized" representation provided by SAS-065. Figure 8.6 provides the results obtained for each of the six experiments. Generally, the loci of positions observed for these approaches form distinct clusters that are in roughly the same relative positions as the idealized representation (Figure 8.5). That is, as the approaches become more network enabled, they move farther away from the origin, up and to the right in the Approach Space. Furthermore, at least for these experiments, there is little overlap between and among the regions. These empirical results confirm that different approaches are indeed located in different regions of the Approach Space.

H3: Traditional hierarchies are located near one corner of the Approach Space, and fully connected edge organizations are located near the opposite corner.

Figure 8.6 also provides evidence that supports the statement that traditional hierarchies (de-conflicted C2 Approaches, in the language of collective endeavors*) tend to be located near one corner of this space, while fully connected edge organizations are located near the opposite corner.

H4: There is no universal C2 Approach or organizational design that is appropriate for all problems (tasks) and circumstances.

8.3.1 Endeavor Space

Mission objectives, the tasks that need to be accomplished, and the circumstances under which these tasks need to be carried out, differ widely. The set of requirements and conditions of interest form an Endeavor Space.† One can think of an Endeavor Space as containing the set of possible challenges or situations that could be faced, just as the C2 Approach Space contains the set of possible C2 Approaches. While the regions of the C2 Approach Space are associated with different C2 Approaches, Endeavor Space regions are associated with different mission archetypes.‡ While in the case of the C2 Approach Space there are a relatively small number of regions of interest (those that correspond to the different C2 Approaches), the topology of Endeavor Spaces has yet to be systematically mapped. In fact, there has not been enough work done on the organization of an Endeavor Space for domain experts (military, business, societal, or government) even to settle on an appropriate number of dimensions, let alone work on scales for each of the dimensions.

---

* Traditional hierarchies are referred to by the SAS-065/085 teams as de-conflicted because they are based upon a traditional Industrial Age model that creates stovepipes that interact with one another only to the extent necessary to de-conflict their activities.

† Alberts (2011), pp. 351–352.

‡ The term "archetype" is used here in the Platonic sense, referring to a pure form that embodies the fundamental characteristics of a thing.

Nevertheless, Endeavor Spaces are critical because we need a systematic way of exploring the relationship between enterprise C2 Approaches, mission archetypes, and the probability of success. In *The Agility Advantage*,* a five-dimensional Endeavor Space was constructed. These five dimensions included type of mission, mission requirements (a combination of a need for timeliness and a need for shared awareness), cognitive complexity, information quality, and infostructure characteristics and performance. The first two involve the nature of the job to be done while the last three involve the conditions under which it needs to be accomplished. With four mission types, nine different combinations of requirements, and three levels of each of the remaining dimensions (cognitive complexity, levels of information quality, infostructure performance), there are a total of 972 distinct "endeavors." While one can certainly think of additional dimensions that may distinguish one endeavor from another, this is arguably sufficient as a point of departure to explore agility in general, and to see if there is any support to the various agility-related assertions and conclusions we have made.

8.3.2 Endeavor Space Applied to Testing That There Is No Universal C2 Approach

We employ an Endeavor Space to test the proposition that there is no universal C2 Approach. In this case, we need only look at a portion of the 972-cell Endeavor Space defined above to come to the conclusion that this proposition is supported by the evidence. Figure 8.7 looks at experimental results that involve four different C2 Approaches, and the nine different mission requirement combinations under three different levels of noise, for one type of mission under favorable conditions (low cognitive complexity and no network damage).† A cell is filled in if the C2 Approach can operate successfully given the mission and circumstances represented by this location in Endeavor Space. There are numerous instances in which more than one C2 Approach can operate successfully in a given region of the Endeavor

---

* Alberts (2011).
† Figure 8.7 is based upon Figures V-26 thru V-29 in Alberts (2011).

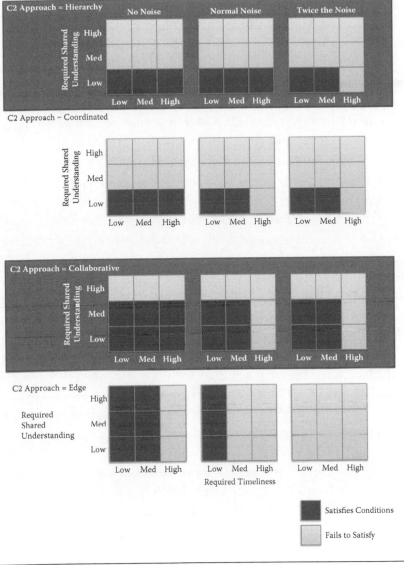

Figure 8.7 Feasibility of various C2 Approaches under various conditions. (Based on Figures V-26 through V-29 in Alberts, 2011.)

Space, but there may be differences in how well and how efficiently they each operate.

Figure 8.7 shows the feasibility of various C2 Approaches under different conditions. The Endeavor Space consists of combinations of mission requirements. The requirement for maximum timeliness can be low, medium, or high; similarly, the requirement for shared understanding can be low, medium, or high; and the noise can be low, medium, or high. All the approaches are feasible when all three are low. There are some circumstances where only a highly decentralized edge approach is feasible, others where only a somewhat decentralized collaborative approach is feasible, and one circumstance where only a centralized hierarchy is feasible. The level of noise has a significant effect on the choice of approach. "Noise" in these experiments refers to intentionally injected irrelevant information. A "no noise" set consists of 34 pieces of information, called "factoids," each of which is useful in accomplishing the task assigned to the organization (a 1:0 SNR). A "normal noise" set is created by adding an equal number (34) of irrelevant factoids to the first set. The irrelevant factoids represent information that could be safely ignored. Thus, this second data set contains a total of 68 factoids and represents a 1:1 SNR (50% noise). A "high noise" condition is simulated by adding yet another 34 useless factoids for a total of 102 factoids, one third of which are signal (a 1:2 SNR, or 67% noise).

As can be seen by inspection, the proposition that there is no universal C2 Approach is supported by the evidence. Of the 27 different conditions considered, while there exists a C2 Approach that can operate successfully in 24 of the 27, no single C2 Approach can operate successfully in more than 14 of the 27 cells. It takes a minimum of three different approaches to accomplish this. In eight cases, only one of the four C2 Approaches is a viable choice. The collaborative approach is the only choice in four cases; the edge is the only choice in three cases; the hierarchy is uniquely qualified in one instance.

8.4 Empirical Support for Hypotheses about Traditional Hierarchies

The organizational approach with which we are most familiar is the traditional hierarchy.

Figure 8.8 depicts the form that a traditional hierarchy takes with teams organized to accomplish specific functions or tasks that can be tackled more or less independently. The approach is well suited to an Industrial Age problem that can be decomposed into parts and worked on in relative isolation from one another.

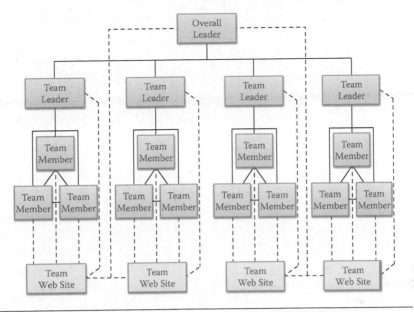

Figure 8.8   Hierarchy relationships and accesses. (From Alberts, 2011.)

H5: Traditional hierarchical organizations are well adapted to some problems and circumstances but are not as well adapted as other organizational approaches for other problems and circumstances.

Hierarchies are well-adapted for particular types of challenges. However, Figure 8.7 shows that even in the case of an Industrial Age challenge (with a low level of required shared understanding), the hierarchy cannot successfully operate under some stresses. In other situations, other C2 Approaches are better adapted than a hierarchy. Figure 8.9 looks at the same region of the Endeavor Space as Figure 8.7, but when there is more than one C2 Approach that can operate successfully, notes the one that is the most efficient.

From Figure 8.7, we note that the hierarchy can operate successfully in eight of the Endeavor Space cells; but from Figure 8.9, we see that it is the best choice in only two of these. Thus, the evidence supports the proposition that hierarchies are well adapted to some problems and circumstances but are not as well adapted as other organizational approaches for other problems and circumstances.

H6: Traditional Hierarchies are often, but not necessarily always, the most appropriate Enterprise Approach for Industrial Age problems.

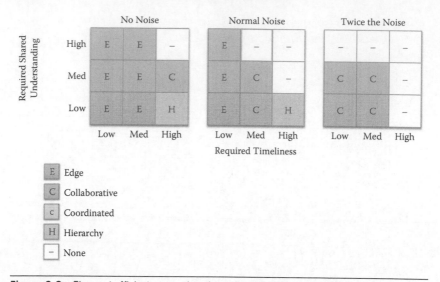

Figure 8.9   The most efficient approach under various conditions. (From Alberts, 2011.)

Looking at Figure 8.9, we can see that the hierarchy does not always represent the best C2 or Enterprise Approach even for Industrial Age problems—that is, problems with a low requirement for shared understanding.

> **H7: Traditional hierarchies do not typically perform well for problems that are complex and dynamic (Big Problems).**

This is discussed more fully in Section 8.5, which explores the proposition that edge approaches are often the best choice for Big Problems. However, we can also find some support for the assertion that hierarchies do not perform well for dynamic and uncertain problems by looking at Figure 8.9. Figure 8.9 shows that for problems that require a high degree of shared understanding and may also be subject to stresses such as noise, hierarchies do not perform well. The high degree of required shared understanding and the presence of noise serve as proxies for complexity and dynamism in these experiments. This *suggests* that hierarchies will not generally be the best choice for Big Problems in the real world, but we must recognize that it is difficult to capture all the dimensions of a "Big Problem" in a set of experiments.

8.5 Empirical Support for Hypotheses about Decentralized, Net-Enabled (Edge) Organizations

Advances in ICT provide not only increased access to information but also the opportunity to rethink how we relate, collaborate, and work together. In short, new ICTs have provided us with an opportunity to create network-enabled organizations. While network-enabled organizations can still allocate decision rights in a variety of ways, the changes in the patterns of interaction and distribution of information that are supported by networking alter their location in the C2 Approach Space. Some of the implications of these changes in location have been widely discussed. For example, the need to resist the increased ability to micromanage that comes with increased ICT has been specifically addressed by the chairman of the U.S. Joint Chiefs of Staff.[*]

Networking capabilities, if properly utilized, can enhance the performance of traditional hierarchies. But they also can create an opportunity to "move to the edge" and create what has been called an "Edge Approach" by distributing decision rights broadly. An Edge C2 Approach is conceptually located in the corner of the C2 Approach Space opposite that of the hierarchy (see Figure 8.5). Figure 8.10 shows the "organization chart" associated with an edge organization. This type of organization can be regarded as something of an abstract end-member along a continuum. Real organizations may not adopt it in pure form, but it is useful in experimentation. Real organizations may, however, assume varying levels of "edge-like" character.

The reasoning behind a move to a more "edge-like" organization is contained in Figure 8.11.

When it was first articulated in the late 1990s,[†] the idea of a self-synchronizing force enabled by shared awareness was controversial. On the one hand, detractors argued it would not work while some proponents overenthusiastically argued that all organizations should move as far to the edge as possible. While this debate raged, there was a growing recognition that there is no universal approach, and that edge, or "edge-like," approaches had a place in the scheme of things.

---

[*] Dempsey (2012).
[†] Alberts et al. (1999); Cebrowski and Gartska (1998); Cebrowski (2003).

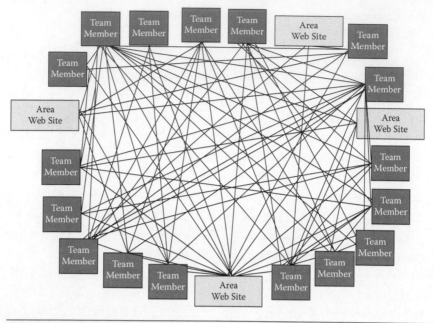

Figure 8.10   Edge relationships and accesses. (From Alberts, 2011.)

H8: Edge organizations are well adapted to some problems and circumstances but are not as well adapted as other Enterprise Approaches for other problems and circumstances.

We concluded above from Figures 8.7 and 8.9 that hierarchies were well adapted for some but not all challenges. We can draw the same conclusion for edge approaches.

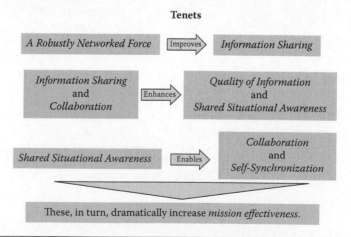

Figure 8.11   Tenets of network-centric warfare. (From Alberts, 2011.)

H9: Edge approaches are often the most appropriate choice for Big Problems.

We have characterized Big Problems as those that are complex and dynamic. For organizations to succeed in meeting the challenges posed by Big Problems, they need to create higher levels of shared awareness *and* be able to respond more rapidly to a constantly changing landscape where the unexpected occurs with increasing frequency. Whether an edge approach is the most appropriate choice depends on the other approach choices that are available, the characteristics of the problem at hand, and the collection of entities required to address it. A key characteristic of Big Problems is that they cannot readily be decomposed into a set of smaller problems that can be solved in isolation. This is because the different aspects of the problem are linked, making the solution of one aspect dependent upon the solution of other aspects. Thus, information sharing and co-evolved solutions are critical.

If we assume that an organization must pick only one approach to C2, then the evidence supports the conclusion that the edge is the most appropriate choice because it is the only choice that can be successful for the most complex problems under the most stressing conditions. NATO Research Group SAS-085, devoted to an exploration of C2 Agility, conducted a series of experiments that examined the relative merits of five different approaches to C2 utilizing a variety of simulation models. They created a pair of maps of the Endeavor Space that support the conclusion that the edge—or, if an organization cannot or will not adopt an edge approach, then the most network-enabled available C2 Approach—is the appropriate choice for Big Problems.

To understand the graphic representations they created, some explanations are necessary. Earlier we discussed the concept of an Endeavor Space. Each of the experiments conducted by SAS-085 created their own Endeavor Space. These Endeavor Spaces were organized so that the challenges became more difficult as they were located closer to the upper right of the space. Figure 8.12 shows the Endeavor Spaces employed in the SAS-085 experiments. Each cell is shaded to correspond to a level of difficulty (how big the problem is): the darker the shade, the bigger is the problem.

For each cell in each of these Endeavor Spaces, the group used experimental results to determine whether or not the C2 Approaches

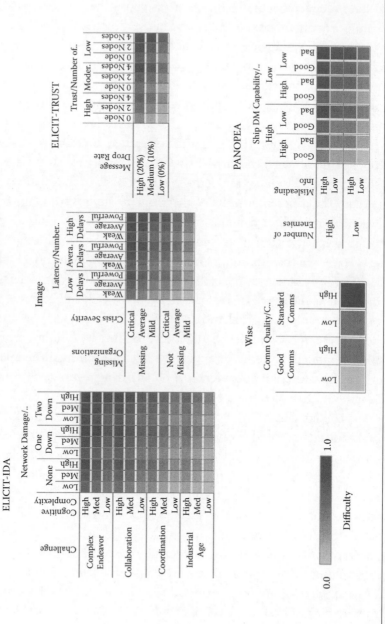

Figure 8.12 NATO SAS-085 Endeavor Space difficulty. (From NATO, 2013.)

simulated would be successful. Based upon these findings the group created two maps. In the first map they assigned, to each cell representing a combination of mission and conditions, the most network-enabled successful approach. Thus, if an edge approach and a hierarchy were both successful, the researchers would associate the cell with the edge approach. In the second map, the researchers used the *least* network-enabled approach as the tiebreaker. Thus, they would associate the same cell in the second map with the hierarchical approach. Comparing these two maps offers us an opportunity to see the most appropriate approach for different levels of problem difficulty. Figures 8.13 through 8.17 present the results from the SAS-085 experiments* (Figure 8.13 for ELICIT-IDA, Figure 8.14 for IMAGE, Figure 8.15 for PANOPEA, Figure 8.16 for ELICIT-Trust, and Figure 8.17 for WISE).

ELICIT-IDA simulated four C2 Approaches with the edge being the most network enabled, and the hierarchy the least. The four-dimensional Endeavor Space employed in ELICIT-IDA consists of 108 cells. As we can see in Figure 8.13, the edge approach dominates the upper right of the Endeavor Space. The fact that it is present in both maps shows that it is the only approach that satisfies these conditions. When the problems are not so big, one can see from the map on the right (with the least network enabled as the tiebreaker) that the coordinated approach can be used for some situations, and when the problems are relatively small, the hierarchy† is appropriate.

IMAGE also simulated four approaches to C2. However, unlike ELICIT-IDA, IMAGE did not simulate an edge approach. Its most network-enabled approach was the collaborative one. The four-dimensional Endeavor Space employed by IMAGE consisted of 54 cells. IMAGE results are presented in Figure 8.14.

The IMAGE map on the left, with the tiebreaker being the most network-enabled approach, shows that the most network-enabled approach that was considered, a collaborative approach, was successful

---

* The original figures (6.13, 6.14, and 6.15) produced by SAS-085 can be found in NATO (2013).
† Since SAS-085 assumes a collective endeavor (e.g., a coalition), it uses the term *de-conflicted* instead of *hierarchy*. In hierarchies the leader's job is to de-conflict the efforts across the silos.

Figure 8.13   Experimental results: ELICIT-IDA. (From NATO, 2013.)

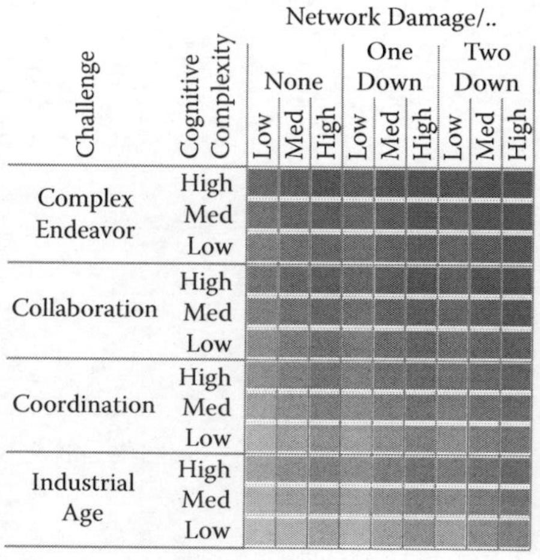

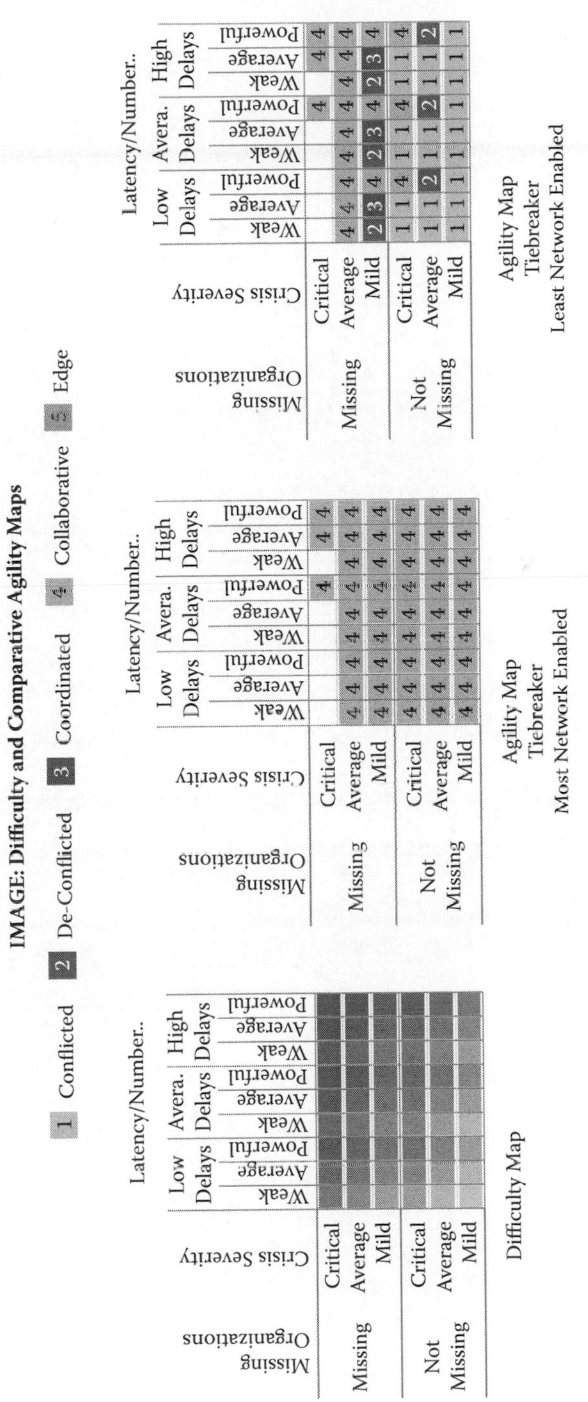

IMAGE: Difficulty and Comparative Agility Maps

1 Conflicted   2 De-Conflicted   3 Coordinated   4 Collaborative   Edge

Figure 8.14 Experimental results: IMAGE. (From NATO, 2013.)

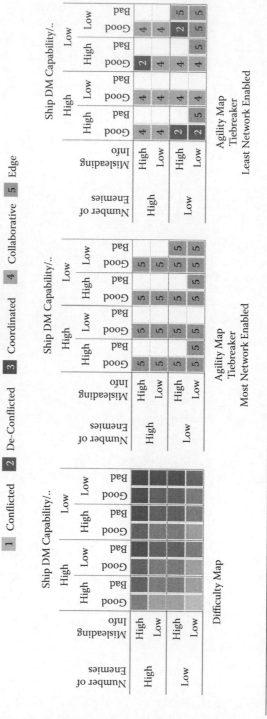

Figure 8.15 Experimental results: PANOPEA. (From NATO, 2013.)

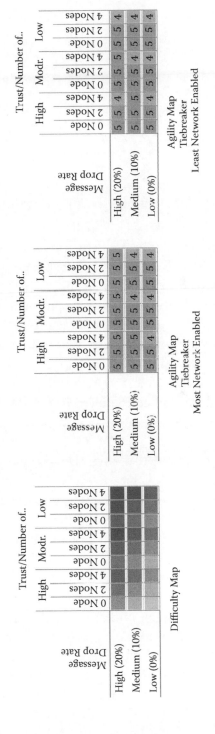

Figure 8.16 Experimental results: ELICIT-Trust. (From NATO, 2013.)

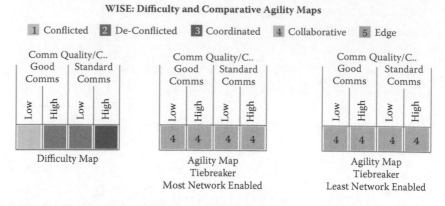

Figure 8.17   Experimental results: WISE. (From NATO, 2013.)

virtually everywhere in the Endeavor Space. The IMAGE map on the right shows that only the collaborative approach satisfies the demands posed by the bigger problems. As the problems get progressively easier, the less-network–enabled approaches become appropriate. The correlation between problem difficulty and the requirement for a network-enabled approach seems clear from these results.

PANOPEA simulated three C2 Approaches including the hierarchy, the collaborative approach, and the edge. The five-dimensional Endeavor Space employed in the PANOPEA experiments consisted of 32 cells. Figure 8.15 provides the results of these experiments.

The results for PANOPEA generally mirror those of ELICIT-IDA and IMAGE even though the mission and circumstances were quite different. Success is possible in 20 of the 32 situations. The edge approach can be successfully employed in all 20 situations but is only essential in five, as less network-enabled approaches can be successfully employed for less difficult challenges.

ELICIT-Trust simulated five C2 Approaches with the edge being the most network enabled. The results from this experiment are shown in Figure 8.16.

In the ELICIT-Trust experiments only the two most network-enabled approaches can meet the challenge. The collaborative approach suffices in 7 of the 27 situations; but some of these 7 are among the most difficult situations faced. Nevertheless, the edge can be successful in 22 of these situations compared to just 7 for the collaborative approach. These results differ a bit from the other experiments because the collaborative approach is not successful in the least stressful cases as one might expect, while the edge approach is.

The WISE experiment simulated just two approaches, the de-conflicted (hierarchy) and the collaborative. The two-dimensional Endeavor Space employed in the WISE experiments consisted of just four cells. Figure 8.17 presents the results of these experiments.

In the WISE experiment, only the more network enabled of the two approaches simulated was successful.

Each of the SAS-085 experiments created its own Endeavor Space. The spaces differed in size and in dimensionality (the number of variables and the range of values they could take on). They also differed in the degree and nature of the stresses that they posed for entities. In some cases, all of the entities simulated were able to operate successfully in some or all of their Endeavor Space. In other cases, there were regions of the Endeavor Space in which none of the C2 Approaches could operate successfully. The proposition *"Edge organizations are often the most appropriate choice for "Big Problems,"* seems to be supported by an inspection of results from these experiments. To determine if this conclusion is justified, the success rate of an edge when faced with the most difficult problems was compared to the success rates for less network-enabled approaches.

Edge approaches were simulated in three of the experiments (ELICIT-IDA, ELICIT-Trust, and PANOPEA). A total of 55 situations were considered to represent Big Problems. They had a combined success rate of 0.564 compared to a combined success rate for all other (less network-enabled) approaches of 0.091. This result supports the conclusion that edge organizations are the most appropriate choice for Big Problems.

H10: More network-enabled approaches are more appropriate for Big Problems than less network-enabled approaches.

Given that not all organizations can, or are inclined to, adopt something close to an edge approach, a more general proposition was tested. The proposition *"More network-enabled approaches are more appropriate for Big Problems than less network-enabled approaches"* was examined by undertaking an analysis of success rates as a function of both problem difficulty and the degree to which the approach was network enabled. In this analysis, data from three of the five main experimental designs were employed. This is because in two of the five, there were no instances of a less network-enabled approach being successful. In WISE, only the collaborative approach was successful. In ELICIT-Trust, only the collaborative or edge approaches were successful. In ELICIT-IDA, IMAGE, and PANOPEA a de-conflicted approach (traditional hierarchy) was successful in some regions of the Endeavor Space, and in the case of IMAGE, even a conflicted (very strictly hierarchical) approach proved to be successful in some cases. In all, this encompassed 196 different cells over three Endeavor Spaces. In 28 of these cases, no C2 Approach proved to be successful. Figure 8.18 compares the success rates for the least network-enabled and the most network-enabled approaches as a function of difficulty. Situations were grouped into those that were most difficult and those that were least difficult, putting aside those in the middle range of difficulty. Approaches were similarly grouped, with conflicted and de-conflicted being categorized as less network enabled while collaborative and edge were considered to be more network enabled.

Given that the success rate for the least network-enabled approaches, when faced with the most difficult problems, is zero, as compared to a rate of 0.619 for the most network-enabled approaches, there is little question that the findings from these experiments support the proposition that "more network-enabled approaches

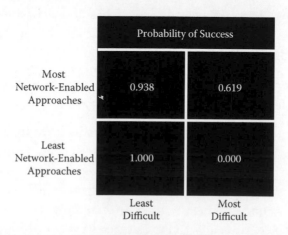

Figure 8.18　Success as a function of approach and difficulty.

are more appropriate for Big Problems than less network-enabled approaches."

8.6 Empirical Support for Hypotheses about the Properties of C2 Approaches

The concept of an Approach Space is not meaningful or useful unless differences in location translate into differences in approach capabilities and performance. Specifically, this refers to how well the approaches located in various regions of the Approach Space perform in different missions and under different circumstances. Thus, the relative locations of a pair of approach options would allow us to determine which of the two was more appropriate for a given situation or which one would be more agile. In this section we consider two propositions. The first involves the relative positions of the regions associated with the different C2 Approaches. The second concerns the location of a particular instantiation of a C2 Approach within the region associated with that approach.

H11: Approaches that are more network enabled also tend to be more agile.

We previously compared the least and most network-enabled approaches (hierarchy and edge) with respect to the types of situations for which they were most appropriate. We now turn our attention away from the abilities of approaches to handle specific situations to the question of their relative agility. An agile approach in this context is one that is able to operate successfully across the Endeavor Space. The greater the part of the space that an approach can "cover," the more agile it is. Thus, in measuring agility, one does not consider whether or not a particular approach is the best approach for a given situation, but rather the total number of cells in which an approach can operate successfully. A simple agility metric (Agility Score) is the percentage of the total volume (in the Endeavor Space) where an approach can operate successfully.

As an approach "moves" away from the origin of the Approach Space, it becomes more network enabled. In this section, we use two ways of looking at the degree to which an approach is network enabled. The first is simply to compare the relative agility of the NATO NEC C2 Approach options (an ordinal scale); the

Experiment / C2 Approach	ELICIT-IDA	ELICIT-TRUST	abELICIT	IMAGE	WISE	PANOPEA	LS-Mean
Conflicted C2		0.04		0.39			0.09 (0.10)
De-Conflicted C2	0.06	0.06		0.50	0.21	0.13	0.14 (0.09)
Coordinated C2	0.10	0.06	0.02	0.54			0.20 (0.09)
Collaborative C2	0.26	0.18	0.13	0.89	0.42	0.47	0.39 (0.09)
Edge C2	0.55	0.46	0.33			0.63	0.59 (0.09)

Figure 8.19　C2 Approach agility scores. (From NATO, 2013.)

second is to actually calculate the distance from the origin for the centroids of each approach (a ratio scale) to see if this distance is related to agility.

Figure 8.19 compares the agility scores obtained for a set of progressively more network-enabled approaches in six different experiments conducted by NATO SAS-085 members.

It should be noted that in every one of these experiments, the agility scores increased monotonically with the degree to which approaches were network enabled. Furthermore, the gains in agility were non-linear and increased as the approaches became more network enabled (see Figure 8.20). An R-squared score was calculated to determine the percentage of variation in agility scores that were explained by the degree to which the approaches were network enabled. The value obtained (0.9937) indicates that this one variable explains virtually all of the variation in agility scores observed in these experiments.

Figure 8.21 shows the relationship between how far from the origin an approach is located and its agility. The coefficient of determination value ($R^2 = .989$) indicates a very strong relationship between these two variables.

Agility is a hedge against uncertainty and, to the degree that one cannot predict the future, the adoption of the most agile approach serves to minimize the risks associated with adopting an inappropriate approach.

While the evidence shows that the most agile approach is also the preferred approach in a large region of the Endeavor Space, this

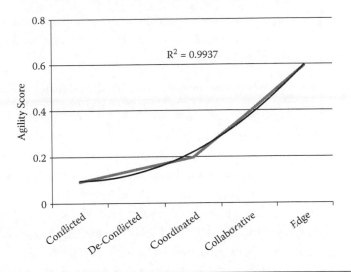

Figure 8.20   Average C2 Approach agility scores. (From NATO, 2013.)

need not have been the case. In theory, even if everywhere in an Endeavor Space there is always another approach that would outperform Approach A, Approach A could still be the most agile of all the approach options. In these experiments, the most network-enabled approach is not only the most agile, but it is also the only approach

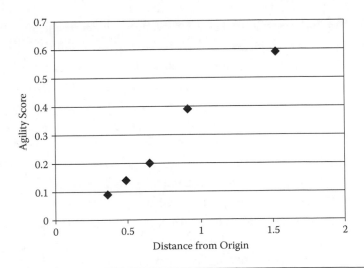

Figure 8.21   Distance from origin versus agility score. (From NATO, 2013.)

that can be successful for some Big Problems. *This suggests that the most network-enabled approach available to an entity is likely to be the best choice for a complex and uncertain world.*

H12: Balanced approaches are more effective than unbalanced ones.

Balance is an important approach attribute because, based upon the empirical evidence, it is associated with effectiveness as well as agility. Before the question of whether or not there is empirical evidence that supports the existence of relationships between approach balance and effectiveness, and between approach balance and agility, can be addressed, approach balance needs to be operationally defined. In the context of a C2 Approach, the variables that need to be "in balance" are those that determine where a particular instantiation of a C2 Approach is located within the Approach Space. We make a distinction here between the various instantiations of, say, a hierarchy, and the region in the Approach Space occupied by Hierarchical Approaches. Thus, in this discussion, an approach corresponds to a single point. This will allow us to identify regions in the C2 Approach Space that contain "balanced" approaches and, by implication, where unbalanced approaches are located. We begin by addressing the question of how to make a determination that a particular instantiation of an approach is balanced. Then we can determine the locations in the Approach Space that are occupied by balanced approaches and see how these locations correspond to performance, to test the hypothesis related to approach balance.

Understanding what makes an approach balanced requires that we understand the role that each of the three dimensions of the Approach Space plays in the ability of an approach to perform successfully, and the interdependencies between and among these dimensions. We cannot treat these dimensions independently because although the C2 Approach Space is depicted as a cube, it is widely understood that these dimensions are not, in theory or reality, orthogonal. These interdependencies are reflected in the ways organizations (approaches) are logically designed as well as in the ways these designs are instantiated and how the corresponding entities actually behave in practice.

A basic organizational design principle is that decision rights should be allocated to individuals who have the wherewithal to make informed and timely decisions. That is, decision makers should not only have the expertise and experience needed for the decisions they are called upon to make, but also have timely access to the information needed. Determinations regarding the allocation of decision rights need to be accompanied by a set of policies, practices, and processes that are designed to support the designated decision makers. Who makes which decisions will, in large part, determine with whom various individuals may interact, and this, in turn, will impact the dissemination of information. Thus, properly designed approaches are, at least in theory, balanced with respect to the three dimensions of the C2 Approach Space.* Being balanced does not guarantee success, but it avoids some of the known drags on performance that can occur when approaches are unbalanced. Thus, all things being equal, a balanced approach should be more effective and efficient than an unbalanced one.

The graphical depictions of different archetypical C2 Approaches (see Figure 8.5) have placed these designed approaches along a diagonal of the cube. This placement along the diagonal was deliberate to convey the idea each of these C2 Approaches are, at least in theory, balanced. This implies that the C2 Approaches that are "off diagonal" will be "less balanced." However, to systematically explore the question of balance, one must move beyond a graphic representation of C2 Approaches in the Approach Space to quantitative measures of each of the three dimensions. Quantifying these dimensions of the Approach Space allows one to locate more accurately a particular instance of an approach as described in theory or as it plays out in practice.

Since the metrics used for the dimensions are somewhat arbitrary, balance does not automatically equate to being located on a straight line diagonal from the traditional hierarchy corner to the edge corner, as is often portrayed. To explore propositions related to approach balance, one must first identify and locate balanced approaches in the C2 Approach Space as well as unbalanced approaches. Alberts et al. (2013) looked for the existence of a "sweet spot" in the C2 Approach Space, a region or regions that corresponded to balanced approaches. Thus, if one considers the region associated with hierarchies, a subset

---

* The concept of co-evolution is related to balance. Co-evolution refers to the adjustments and adaptations that are made over time between and among the elements of DOTMLP (Doctrine, Organization, Training, Material, Leadership, and Facilities) to take advantage of an entity's capabilities. It was introduced into the literature in connection with the desire to leverage the power of networking that became known as network-centric warfare (see Alberts et al., 1999, p. 197).

The Scales Employed

C2 Approach Dimension	Nature of Measure	Metric
Allocation of Decision Rights (ADR)	Degree to which decision rights are distributed; a measure of participation in decision making	Ratio of the number of individuals exercising decision rights to the total number of individuals
Patterns of Interaction (PoI)	Density of interactions between and among individuals; a measure of quality, frequency, and reach of the interactions	Normalized square root of the number of information-related transactions
Distribution of Information (DoI)	Degree to which individuals have access to available information	Average percent of available factoids received by an individual

Figure 8.22 C2 Approach dimension scales. (Adapted from Alberts et al., 2013.)

of this region that would be occupied by balanced hierarchies would, at least in theory, have a competitive advantage. The same would be true for each of the other approaches. Taken together, these subregions would form the sweet spot in the Approach Space.

The first step is to locate a given C2 Approach in the Approach Space based upon observable, quantified metrics for each of the dimensions. Quoting Alberts et al. (2013):

> There are, of course, many ways that one can think of to scale each of the approach space dimensions. Furthermore, given the multi-dimensional aspects of each of these dimensions, it is unlikely that a simple scale (consisting of one variable) will be able to capture all of the factors of interest. For example, not all decisions, interactions, and pieces of information are of equal value and thus any scale that simply "counts" these will not capture important nuances. While we can expect that we will find better ways to measure each of the dimensions of the C2 Approach Space as we gain experience, we needed to start somewhere.[*]

Alberts et al. (2013) defined three dimensions as shown in Figure 8.22. Note that the metric chosen for Patterns of Interaction (PoI) is not a linear function of the number of interactions. This is because the number of interactions does not increase linearly as the C2 Approaches become network enabled. The Distribution of Information (DoI) metric, unlike the PoI metric, is not a process

---

[*] Alberts et al. (2013).

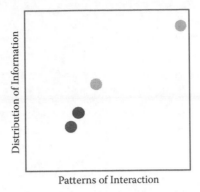

Figure 8.23   Intended approach locations in the C2 Approach Space.

measure but an outcome measure: how much of the information ulti-mately is distributed, on average.

Since the allocation of decision rights (ADR) is fixed for each approach in these experiments, we need only look at PoI and DoI to see the size and shape of the regions where the different approaches are located and how the locations are affected by circumstances. Figure 8.23 shows the locations implied by design based upon the descriptions of the four NATO SAS-065 networked-enabled C2 Approaches. These intended locations, while progressing from one corner to another, do not lie strictly along the diagonal of the cube depicted in Figure 8.5, but along a curve. Upon reflection, this makes sense given the nonlinearity of the PoI metric. Since these loca-tions are theoretical, they may not reflect how such organizations will behave in practice. While this may be so for unexpected or very stressful circumstances, one cannot dismiss the possibility that even under normal or expected circumstances the approach in practice will be located at some distance from its intended (design) position.

To understand how organizations that adopt these approaches might behave in practice, a number of different approaches were instanti-ated in a simulation.* Figure 8.24 depicts the observed locations in the Approach Space that correspond to specific simulation runs (each run employed a different set of conditions) for each of four differ-ent approaches. Figure 8.24 portrays information about the different

---

* The simulation employed the ELICIT environment and was reported by Alberts et al. (2010).

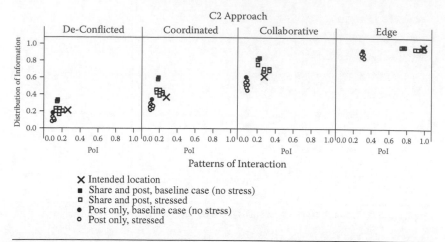

Figure 8.24   Observed versus intended approach locations. (Adapted from Alberts et al., 2013.)

approaches to C2, how they behave under different conditions, and the effect of the information-sharing policy. Squares are used for runs where individuals shared information redundantly, by both sharing and posting. Circles are used for simulation runs where a "post-only" policy was employed. The intended positions for each approach (as depicted in Figure 8.23) are designated by an "x". The baseline cases are filled, while other circumstances (the presence of one or more stresses such as noise or a heavy workload) are unfilled. Figure 8.24 enables us to make the following observations.

First, the size of the regions defined by the locus of points associated with a type of C2 Approach (e.g., hierarchy, coordinated, collaborative, edge) grows bigger as the approach becomes more network enabled.

Second, this is a result of the growing separation into two distinct clusters (circles versus squares) attributable to the two information-sharing practices. This separation is most pronounced for the edge. This result makes sense since far fewer transactions are generated by a post-only practice, and this has some effect on the distribution of information, particularly for less-network–enabled approaches. Thus, information-sharing policy is a key variable, one that accounts for the relative positions of approach instantiations by acting as a constraint that affects the observed patterns of interaction and distribution of information.

Third, the intended positions (marked by an "x"), while they are located in the general vicinity of the approach clusters, are certainly not centered in the regions observed. Except in the case of the collaborative approach, the intended positions reflect both a greater number of inter-actions and less dissemination of information than actually occurred.

Fourth, the baseline cases, those that are subject to the least stress-ful conditions, are consistently located on the upper or upper-left edge of their respective clusters. We can conclude from this that, in prac-tice, when entities are subjected to one or more stresses (a need to deal with bigger problems), their observed positions in the Approach Space shift down and to the right.

At this point, all we have are relative locations in the Approach Space and thus, not enough information to determine which of these locations correspond to balanced approaches. There are two ways to determine balance. The first is to perform a forensic analysis to see if disconnects and inefficiencies that adversely affected entity perfor-mance could be traced to imbalances between and among the dimen-sions of the C2 Approach Space. While this can and should be done both to advance our understanding of the behaviors and performance of different approaches, it is a time-consuming task that requires detailed instrumentation of reality or detailed analysis of simulation transaction logs. A systematic forensic analysis was not performed for these experiments.

Given the rationale for balance, one can conclude that unbalanced approaches would be at a competitive disadvantage when compared with balanced approaches—that is, their effectiveness and efficiency would be reduced. Thus, we could look instead at the locations that resulted in the best performance and the worst performance. Since, in the simulation, the values of the other independent variables were held constant, we can attribute these variations in performance to the differences in locations within an approach region. Figure 8.25 desig-nates the locations of the best and worst performers for each approach and information-sharing policy.

These results consistently show that regardless of the NATO Network-Enabled Capability C2 Approach and the policy variant, the best performers are those that are more efficient than the worst performers. That is, they are located higher and farther to the left. This translates into more information dissemination (higher) with fewer transactions (to the left). If one looks again at Figure 8.24, the locations of the best performers correspond to the baseline cases

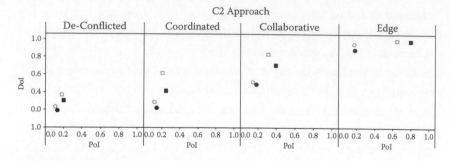

Figure 8.25 Best (unfilled symbols) and worst (filled symbols) performers for each approach and information-sharing policy (DoI, Distribution of Information; PoI, Patterns of Interaction). (Adapted from Alberts et al., 2013.)

where conditions are the least stressful. This leads to the conclusions that the locations associated with the baseline cases are the most balanced while those that are farthest away from these locations (within a cluster) are the most unbalanced. That is, *balance* and *competitive advantage* are concepts that are inseparable in this context.

To continue with their analysis of balance and its relationship to effectiveness (desirable outcomes), SAS-085 used the locations of the baseline cases for the post-and-share policy to create an empirically defined diagonal. They then calculated the distance from this diagonal and compared the results observed from those on or close to the diagonal with those at a distance from the diagonal. Figure 8.26

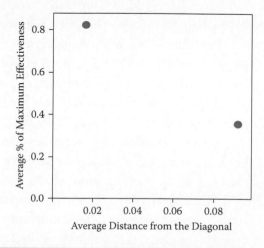

Figure 8.26 On- versus off-diagonal performance. (From NATO, 2013.)

compares the performance of approaches on or close to the calculated diagonal from those that are farther away from the diagonal.

The on-diagonal approaches performed more than twice as well as off-diagonal approaches. In theory, the approach options available to an entity have been designed to be in balance. In practice, balance is a function of whether the assumptions made by approach architects are valid for the mission and circumstances at hand. Success often depends on being able to balance a need to do the right thing and the need to do it in a timely manner. This, in turn, requires timely access to information. Whether or not the designated decision makers will actually be able to obtain the information they need when they need it, depends upon the requirements of the mission as well as the prevailing circumstances. These circumstances include the performance of ICT. Other assumptions include how people will behave. This in turn depends upon the expertise, experience, and cognitive capabilities of individuals and the levels of trust they have in each other and in the systems that support them. These and other variables influence the actual location of an approach in the Approach Space. The experiments discussed above show that mission challenges and adverse circumstances can move the location of an approach, throwing it into an unbalanced state that in turn can adversely affect its performance. Thus, an approach that could be successful, if executed as intended, can fail if it is not implemented as designed. Therefore, entities need to monitor how their adopted approaches are carried out in practice and take appropriate actions to ensure balance.

8.7 Empirical Support for Hypotheses about Organizational Design Choices Being Constrained by Capabilities and Circumstances

A basic theme of this book is that success depends, in large measure, on the appropriateness of the approach option selected. Selecting the most appropriate approach begins, but does not end, with an understanding of the nature of the mission to be accomplished. In particular, the complexity and dynamics of the situation and the attendant risks must be understood. While on paper an entity can choose from among any of the approach options in its tool kit, circumstances can constrain this choice. We have previously seen how effectiveness can be degraded by a lack of balance. Furthermore, we have seen experimental results that show that various stresses can move one's location in the Approach Space from a region of balance to one of imbalance. In this section, we look at specific circumstances and stresses that need to be factored into a selection of an approach. The first circumstantial factor we consider is the quality of information that is available to the enterprise.

H13: The quality of the available information limits the potential effectiveness and efficiency of an enterprise.

While decision makers' experience and expertise can sometimes compensate for low information quality, such low information quality generally degrades the quality of decisions. This affects mission performance. In this section, we explore empirical evidence that quantifies this relationship. We examine the results of a set of experiments that can vary aspects of the initial quality of the information available to the enterprise, and observe the resulting information-related behaviors that affect enterprise information quality over time.

A great variety of experiments could be designed to investigate the relationship between information quality and mission performance. For example, the completeness, correctness, consistency, timeliness, relevance, accessibility, and ease of use of the information available can be dynamically manipulated. There are also variables that impact information-related behaviors, such as organization structure, policies, and processes. These variables can impact measures of enterprise information quality, including those that reflect aggregate information transactions and information dissemination.

In the ELICIT experiments whose results we discuss here, enterprises were given all the information necessary to accomplish their assigned tasks. Thus, the data set was always complete (i.e., the completeness metric was 100%), and all individual pieces of information were accurate (truth and believability were 100%).

However, the individual pieces of information differed with respect to their relevance. Some were relevant (useful), and others were not. The ratio of relevant to irrelevant information, or what the researchers called the *signal-to-noise ratio*, differed across sets of simulation runs. Since all data sets used were complete, signal-to-noise ratios were varied by injecting "noise"—that is, by adding irrelevant pieces of information, as in Section 8.3.2. *Enterprise signal-to-noise ratio* is an example of an enterprise information quality metric that while it can be set at the beginning of an experiment can also be varied over time.

Other enterprise information quality measures can be influenced but not set as they are a result of emergent behaviors. For example, human (or in this case agent) behavior can either amplify the noise or reduce it. Both the initial value of information quality and how it varies over time have an impact upon task performance and mission success.

The values of these variables can provide explanations for experimental outcomes. Take, for example, an enterprise information quality measure that reflects the amount of information that flows throughout a system or an entity. This enterprise measure can make a significant difference in performance. This is because *enterprise timeliness*, expressed as a fitness-for-use quality metric, can be significantly impacted by the nature and size of information flows. Given that networked organizations seek to increase shared awareness, the extent to which information is shared is also of considerable interest, and *enterprise shared information* is another enterprise information quality metric.

The hypothesis considered here (quality of the available information limits the potential effectiveness and efficiency of an enterprise) is widely believed to have merit, as evidenced by the focus on information requirements and systems. A general belief in the importance of information quality, however, is not the same as the ability to quantify the value of a particular improvement in information quality. Decision theorists have developed ways to calculate the value of information including the value of perfect information, but only for specific decision problems under given sets of circumstances. Recent work on C2 Agility[*] provides a basis for generalizing this idea using enterprise measures of information quality. The ELICIT experimental

---

[*] Alberts (2011).

results discussed in this section illustrate the relationship between an enterprise information quality measure (the signal-to-noise ratio of the information available to the enterprise) and two aspects of task performance.*

The first aspect of task performance measured in these experiments is *enterprise responsiveness*, defined as a measure of the time it takes to develop the correct solution relative to the time that is available. In cases where multiple individuals arrive at or become aware of the correct solution, responsiveness is calculated using the time when the correct solution was first developed. Responsiveness is zero if the time it takes to develop a correct solution is greater than the time that is available (a failure). Responsiveness is equal to "1" if the task is performed instantaneously such that all of the available time remains. A value between 0 and 1 reflects the percent of available time remaining. The second aspect of task performance is a measure of shared understanding, where understanding equates to the development of a correct solution to the problem at hand, and enterprise shared understanding to the percentage of enterprise participants that get the correct solution.

The experiments discussed here employ both aspects of task performance, since mission requirements can differ. Some missions require a high degree of responsiveness but do not require high levels of shared awareness. For some of these missions, the only thing that matters is that a correct solution is developed. On the other hand, other missions require high levels of shared understanding but do not require a high degree of responsiveness, while still other missions require both in varying amounts. Thus, task performance requirements are a function of the mission.

To explore the impact that enterprise information quality has on task performance, three data sets were constructed that correspond to three signal-to-noise ratios (SNRs). As in Section 8.3.2, a "no noise" set consists of 34 pieces of information, called "factoids," each of which is useful in accomplishing the task assigned to the organization (a 1:0 SNR). A "normal noise" set is created by adding an equal number (34) of irrelevant factoids to the first set. The irrelevant factoids represent information that could be safely ignored. Thus, this second data set

---

* Alberts (2011).

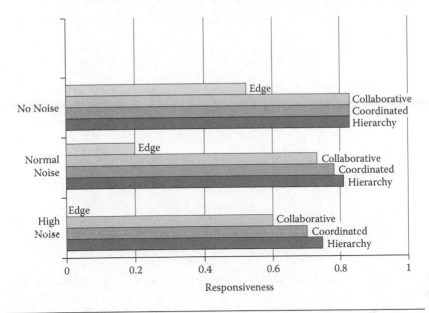

Figure 8.27 Responsiveness as a function of noise and approach. (From Alberts et al., 2012.)

contains a total of 68 factoids and represents a 1:1 SNR or 50% noise. A "high noise" condition is simulated by adding yet another 34 useless factoids for a total of 102 factoids, one-third of which constitute a signal (1:2 SNR or 67% noise).

Figure 8.27 shows the impact of this introduced noise on the responsiveness of various collective C2 Approaches. These results clearly show that noise constrains performance (responsiveness). Further, this is true regardless of which C2 Approach is employed, although the most decentralized (edge) approach is much more affected by high noise levels than the others.

From Figure 8.27 it is clear that the more network-enabled approaches are more adversely affected by increases in noise levels than the less network-enabled approaches. The reason for this is that more networked approaches are designed to share information widely and encourage collaboration in order to achieve higher levels of shared awareness. The collaborative approach develops five times the shared understanding of the hierarchical or coordinated approaches, and the edge approach develops almost 20 times the level of shared understanding. Developing higher levels of shared understanding entails more communication between and among all of the entities in the

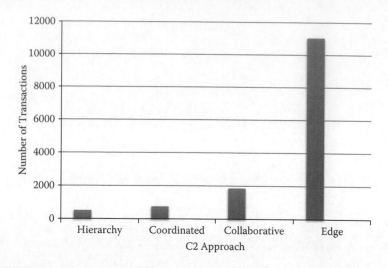

Figure 8.28   Number of information transactions as a function of approach. (From Alberts et al., 2012.)

community. This communication grows exponentially as an enterprise becomes more network enabled, as shown in Figure 8.28.

We next look at another key performance variable, the ability to achieve shared understanding. The requirement comes from the nature of the mission and the need for synchronization. The ability to achieve a given level of shared understanding (something that happens over a period of time) is a function of noise (quality of information), patterns of interaction, and access to information. To determine if an enterprise is able to satisfy mission requirements, we look at the ability of an enterprise to develop a given level of shared awareness within a given period of time employing a set of different C2 Approaches with different levels of noise present.

Figure 8.29 depicts an Endeavor Space with 27 different circumstances, one for each combination of SNRs, the mission requirement for shared understanding, and the mission requirement for maximum timeliness. The feasibility (that is, the ability to complete the mission under the specified conditions) of each of four different collective C2/organizational approaches was determined. All of the approaches are feasible when the SNR is 1:0 (no noise), the mission requirement for shared understanding is low, and the mission requirement for maximum timeliness is not high. There are some circumstances where only an edge approach is feasible,

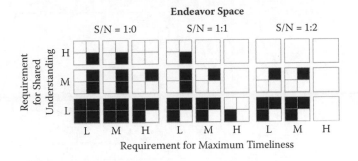

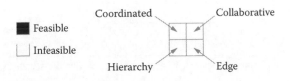

Figure 8.29   Mission success as a function of noise and approach. (From Alberts et al., 2012.)

others where only a collaborative approach is feasible, and one circumstance where only a hierarchy is feasible. Looking at the patterns of feasibility in the regions of Endeavor Space that correspond to different SNRs, it is easy to see that the level of noise significantly influences the choice of approach. An edge approach may be a good choice when information quality is moderate or high, required timeliness is not high, and there is a high requirement for shared understanding. A hierarchical approach may be a good choice when information quality is low or moderate, there is a high timeliness requirement, and there is a low requirement for shared understanding.

Based on these experimental results, enterprise data quality (as measured by the absence of irrelevant information) clearly has an impact on both responsiveness and shared understanding. Hence, there is evidence that supports the hypothesis that information quality can constrain enterprise performance. There is also evidence that an enterprise can offset this constraint by choosing the most appropriate C2 Approach.

H14: Some Enterprise Approaches are better able to take advantage of high-quality information, or perform despite low-quality information, than others.

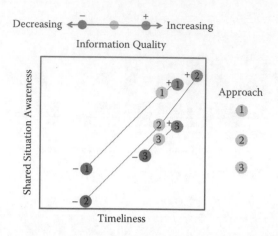

Figure 8.30 Hypothetical situations showing what the evidence of approach sensitivity to noise might look like.

If this statement is true, then we might expect to see some approaches perform relatively better than others as information quality increases, or, conversely, experience larger degradations in performance when the information quality is lower. Figure 8.30 depicts a hypothetical situation illustrating this. In this hypothetical case, Approaches 1 and 2 are very sensitive to degraded information quality while Approach 3 is not. Approach 2 can take better advantage of improved information quality than either Approach 1 or 3.

In the previous section, we used the signal-to-noise ratio as a proxy for information quality. Here, we expand our definition. We refer the reader also to Chapter 5.

The dynamics of information quality and one's approach are inter-related. As a result, some entities are able to maintain or improve the levels of information quality from the set of sources to which they have access, while other entities cannot, and therefore experience degraded information quality. For example, inconsistent data can be resolved if individuals with appropriate knowledge and access can interact with one another. The same is true for incomplete information. Some approaches can deal with noise (lack of relevance) better than others, filtering out irrelevant data and thus reducing workload, reducing confusion, and avoiding associated delays. Some approaches can ensure that information is made accessible to those who require it in a timely manner, while others are less able to do so. In other cases, available information can be buried or delayed resulting in a loss of value. Therefore, when the initial level of quality is high, it seems obvious that some approaches will be in a better position to take advantage of this and create value while others cannot.

Given all of the factors that contribute to information quality, a great deal of experimentation and analysis will be required to explore the interrelationships between C2 Approach and information quality. The following discussion will highlight some of the characteristics associated with specific C2 Approaches that can have a significant impact on either improving or degrading the quality of available information. For this discussion, we take, as a given, the quality of information available to an entity. We consider this to be a function of its sensors and other means of collection. We are interested here in what happens when the information is processed by and flows through an organization (or collection of organizations) to be ultimately accessible to users. To simplify the discussion we will consider only two contrasting approaches (hierarchy and edge).

On the face of it, from an individual's perspective, the two most important measures of information quality are correctness and timeliness. From an enterprise perspective, the degree to which information is shared (measured by accessibility and information in common) is critical, as it enables shared awareness. While the other factors that affect information quality are important, these two are "show-stoppers" when they fail to reach the thresholds necessary to inform decisions. Approaches differ in the way decision rights are allocated, and thus, correctness and timeliness need to be measured at the point(s) of the decision(s). Mission success depends upon all of the decisions that are made, many of which need to be taken collaboratively and need to be synchronized, and hence the need to measure information access and sharing. From our earlier "taxonomy of failures" discussion (Chapter 7), it is clear that a significant number of failures are due to degradation in information quality, exacerbated by the adoption of a particular approach to C2.

Since, in the context of the complexity and dynamics of Big Problems, information is often fragmented in the form of "dots" (partial information coming from a large variety of sources), a major task of C2 is to "connect the dots." This task is made significantly more difficult when the information required involves multiple knowledge domains.

A series of experiments reported by Alberts (2011) compared the ability of different approaches to C2 to "connect the dots" in a timely manner as a function of the mission challenge, timeliness requirements, and circumstances.* The mission challenges differed in the degree to which the required information was fragmented across the domains that corresponded to organizational stovepipes. The results of these experiments show that the hierarchy, except in the case that was characterized as an Industrial Age problem,[†] was unable to connect the dots while the edge was able to do so across the mission spectrum.[‡] This was attributable to a lack of connectivity (and interactions) across the stovepipes that dominate the structure of hierarchies. However, the edge approach was not without its problems. Edge organizations can, if they do not employ a flexible information-sharing policy, become inundated with unnecessary transactions (information sharing); in other words they can experience information overload that adversely impacts cognitive workload and available communications bandwidth and results in data getting lost or delayed (Figure 8.27). Adopting a more efficient information-sharing policy (post only[§]) solves this problem but makes an entity more vulnerable to a degraded communications/information infrastructure. A flexible policy that recognizes, and accounts for, the state of the network is potentially the most agile, as it may be able to take advantage of a benign environment and cope with a degraded one.

The quality of information available to an entity (or enterprise) can be considered to be a baseline and used to compare the performance of an entity employing different C2 Approaches under different conditions.

---

* In these experiments the quality of information available to the organization was complete and current enough for all mission challenges under benign conditions. Thus, failures to succeed were a result of information quality degradation.

[†] An Industrial Age problem in the context of these experiments is one that is amenable to decomposition into domain-specific tasks, each of which can be accomplished independently with information and expertise available to a single stovepipe. As the missions become more complex, the tasks require more interactions between and among stovepipes.

[‡] Alberts (2011), pp. 346–347.

[§] A "post only" policy employs Web sites. Those with information post it, and those in need of information pull it. The U.S. Department of Defense has been encouraging a "post before process" to help ensure that information is available to those who need it in a timely manner, particularly to those the "owners" of information are not aware of.

How this information is disseminated (access) and shared (collaboration) determines what is available to which decision makers. How many decision makers there are, what decisions they can make, and where they are located, are functions of the approach being employed.

In addition to the correctness of the information available to decision makers, we need to know, at a minimum, the currency of the information. While some aspects of information quality can be improved by sharing, consultation, and analysis, the currency of the information cannot. Providing access to information (that includes discovery) takes time. How long it takes for information (or information about information) to get from the boundaries of an entity to the individuals who require it is a function of the C2 Approach (and associated policies and practices). Hierarchies are able to minimize the time delays associated with a subset of the information available for a subset of the decision makers. The patterns of interaction associated with hierarchies result, however, in some information not being disseminated within stovepipes in a timely manner, and, in some cases, not been disseminated at all beyond the stovepipe where it was collected. Edge organizations disseminate more information more widely, but as a result, access to some information may be delayed. Thus, we can see that the information quality experienced by the designated decision makers will be a function not only of the quality available to the enterprise but also of the C2 Approach adopted. Mission requirements ultimately determine whether the quality of information is sufficient.

Figure 8.31[*] presents experimental results that show how the quality of the information available to an enterprise affects the ability of the enterprise to develop shared awareness in a timely manner as a function of the C2 Approach. In these experiments, the edge was the most sensitive to poor information quality. In fact, it failed to satisfy mission requirement when the noise level was high. The edge was also best able to take advantage of higher-quality information. The hierarchy was the least sensitive. It was not greatly impacted by lower information quality nor able to take advantage of improved information quality.

---

[*] Figure 8.31 is based on the results depicted in Figure V-25 in Alberts (2011), p. 350.

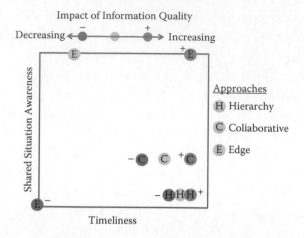

Figure 8.31 Impact of information quality. (Based on results by Alberts, 2011.)

This evidence supports the conclusion that some approaches are better able to (1) take advantage of high-quality information or (2) perform despite lower-quality information than others.

H15: Communications and information system capabilities can limit which approaches are feasible or appropriate.

The ability of individuals to interact and have access to information determines where in the Approach Space an entity is capable of operating. A lack of connectivity prevents individuals from contributing information, accessing nonorganic information, and participating in processes. The capabilities of a tactical network depend on an interaction between design and circumstance. Radios have only so much power, and they have specific waveforms and antenna characteristics. Connectivity depends upon the spatial deployment of units, and conditions that include terrain, weather, and jamming. Available bandwidth is another important consideration. Thus, before comparing the pros and cons of different approach options, there needs to be an analysis that looks at the approach region to see if there are any constraints that make parts of it inaccessible or problematic.

Figure 8.32 is one of a series of graphics that presents the results of experiments that show the locations in the Approach Space when an

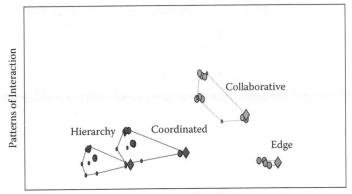

Figure 8.32   Within-cluster location and success. Patterns of Interaction are broader upward, and Distribution of Information is broader rightward. (Adapted from NATO, 2013.)

approach was successful (larger circles or diamonds) and when it was not (bullets). For three of the approaches there are some locations that are associated with failure. Looking at the hierarchy cluster (far left), one can see that there is a threshold in both the PoI and DoI dimensions that needs to be met in order to succeed when employing this approach. If an entity's ICT capabilities are not sufficient to operate above these thresholds in the prevailing circumstances, then for all intents and purposes, the approach is not a viable option.

H16: Trust between and among individuals or organizational components can limit approach options.

To the extent that a lack of trust inhibits the acceptance of information, information sharing, and interactions between and among individuals, it constrains PoI and DoI.

H17: Attacks that degrade communications and information capabilities impact network-enabled approaches more than traditional hierarchies.

Cyberattacks can vary in intent, target, method, and impact. A recent study by Uma and Padmavathi surveys cyberattacks and offers an approach to their classification.* One of the dimensions of

---

* Uma and Padmavathi (2013).

the taxonomy is "Purpose and Motivation." The authors of the study propose the following categories for this dimension:

- Obstruction of information
- Counter-cyber-security measures
- Retardation of decision-making process
- Denial of services
- Abatement of confidence
- Reputation denigration
- Smashing up legal interest

The experiments designed and conducted by Alberts (2011) included limited attacks in which some but not all information flows were obstructed (blocked or delayed). This in turn had the potential to retard dependent decision processes and adversely impact the correctness of decisions. "Limited" in this context refers to a less-than-catastrophic loss of performance—that is, a degree of degradation that adversely impacts the behavior of the organization and impairs its ability to be successful. A limited attack may prevent an entity from functioning successfully in some but not in all missions and circumstances. It may increase the cost of success or it may reduce the measure(s) of effectiveness (even if these still remain in an acceptable range).

Since many of the consequences of a cyberattack can be the result of a variety of other causes, an analysis that focuses on the ability of an approach to cope with the conditions that are created (regardless of the reason) has utility. Thus, to explore the relative ability of various approaches to remain successful in the face of cyberattacks one can simply manipulate the variables that would be impacted by such an attack. For example, by disabling a link, either permanently or temporarily, one can see what would happen in the event of a cyberattack, a power outage, a severed communications link, jamming, or for that matter a system malfunction.

The experiments reported on in Alberts (2011) included looking at a loss of connectivity, specifically the loss of one or two links. The loss of a link in these experiments prevents a pair of individuals from directly communicating with each other. It requires information to flow through other individuals or a Web site. The loss of a Web site is more significant in that it affects all individuals who post to or pull from that Web site. The results of these experiments thus offer

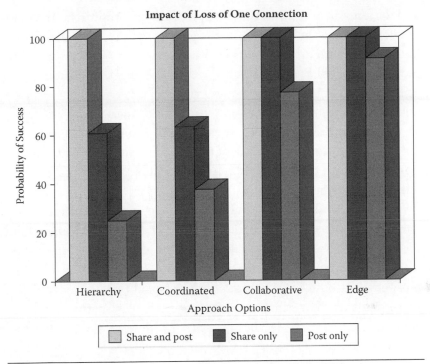

Figure 8.33   Probability of success with one link down. (From Alberts, 2011.)

some evidence of the ability of the different approaches to function successfully in the face of some types of cyberattack, or physical attack on the communications and computing infrastructure.

Figure 8.33* compares the probability of success for different approaches when faced with the loss of a single random link under three information-sharing behaviors. The "Industrial Age" behavior limits the sharing of information to point-to-point communications (share only). That is, one individual shares or pushes information to one individual at a time. "Information Age" behaviors involve the exclusive use of Web sites (post only). Individuals disseminate information by posting information to a Web site and find the information they need by pulling from a Web site. In the third case, individuals use both "share" and "share and post." These different information-sharing practices provide different degrees of redundancy. With increased redundancy come greatly increased traffic flows. The amount of redundancy

---

* From Alberts (2011), Figure V-46, p. 396.

and the amount of traffic are a function of the C2 Approach. The more network enabled the approach is, the greater the inherent redundancy but the larger the transactional load on the system. Given that many tactical systems have limited bandwidth, keeping traffic flows manageable is important. At some point, the traffic overwhelms the available bandwidth, and this prevents timely information dissemination and retards decision making. Thus, the selection of the appropriate approach and information-sharing practice is a function of the situation (damage to the information and communications infrastructure and mission requirements).

These results show that all approaches can, if they adopt a redundant share, post and pull practice, withstand the loss of a single link. However, only the more network-enabled approaches can sustain damage and maintain a high probability of success while adopting the most efficient post-only practice. The price paid (lower probability of success) for efficient behaviors is far higher for less network-enabled approaches. Figure 8.34, when compared to Figure 8.33, shows that

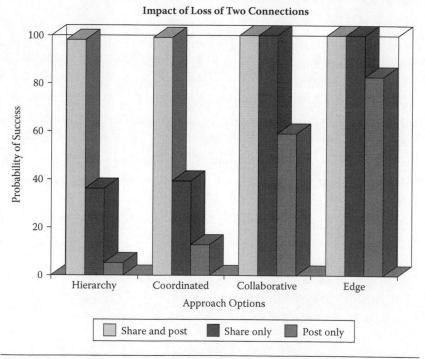

Figure 8.34 Probability of success with two links down. (From Alberts, 2011.)

the greater the damage to the network, the less able the hierarchy is to adopt efficient information dissemination practices.

Thus, in hostile (cyberattacks) and austere (low-bandwidth) environments, only the most network-enabled approaches can successfully operate. Therefore, both logic and the evidence support the conclusion that edge organizations are less adversely impacted by limited cyberattacks than traditional hierarchies.

8.8 Hypothesis about Enterprise/C2 Agility

H18: Changes in circumstances can require a change in Enterprise Approach, and an agile enterprise must be able to execute such a change.

Changes in circumstances present both opportunities and challenges. We will consider two types of change: the first, a change to mission requirements; the second, a change to enterprise communication and information-related capabilities. Both types of changes, for different reasons, may require a change in the Enterprise Approach.

8.8.1 Responses to Change in Mission Requirements

Figure 8.35 shows that the selection of an approach begins with an understanding of the mission requirements and circumstances. This understanding amounts to locating the situation in an Endeavor Space.

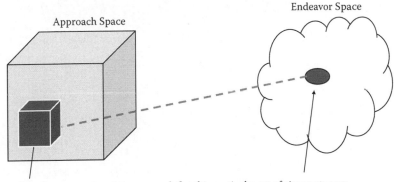

Identifying the Appropriate Enterprise Agility

Endeavor Space

Approach Space

This is a most appropriate approach for this particular set of circumstances

Figure 8.35   Appropriate enterprise approach. (Adapted from NATO, 2013.)

When mission requirement change,
a different approach might be more appropriate

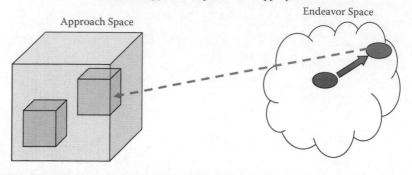

Figure 8.36 Change in enterprise approach due to change in mission requirements. (Adapted from NATO, 2013.)

When mission requirements change, so too does location in the Endeavor Space. If the change is large enough, the mapping to the Approach Space points to a different region in the Approach Space (see Figure 8.36). As a result, the current approach is no longer the appropriate one.

For example, suppose an event occurs that requires a coordinated response. This, in turn, requires that shared awareness among a large set of individuals (entities) be achieved. Figure 8.37 shows that an increase in the need for shared awareness requires a change in approach.

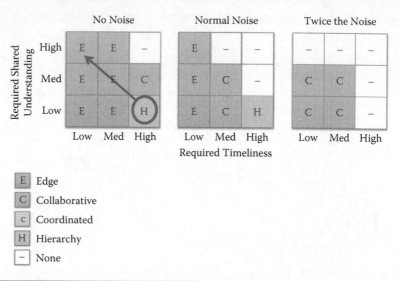

Figure 8.37 Need for increased awareness and approach appropriateness. (Adapted from Alberts, 2011.)

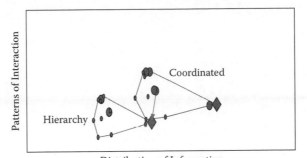

Degraded Communications and Information Infostructure

Figure 8.38 Degraded performance and movement in the Approach Space. Patterns of Interaction are broader upward, and Distribution of Information is broader rightward. (Adapted from NATO, 2013.)

8.8.2 Responses to Change in Communication and Information-Related Capabilities and Behaviors

The location of an approach in the Approach Space is determined by the actual patterns of interaction and the actual distribution of information that is achieved. These in turn depend upon the performance of communications and information infrastructure (infostructure). The dynamics of the battlefield can change, resulting in forces becoming more dispersed geographically, or moving into forested or mountainous terrain, adversely affecting line-of-sight communications. The performance of the communications and information systems can be adversely impacted by internal problems or external attacks (kinetic or cyber). Reduced connectivity and bandwidth, for whatever reason, may have an adverse impact on information sharing. As a result, the location in the Approach Space will shift down and left in Figure 8.38. This makes execution of the approach to the right (in this case a coordinated approach) unbalanced and, based upon these experimental results, unlikely to succeed. However, the new position in the Approach Space corresponds to a point in the hierarchical approach region that is likely to succeed.

The need to change one's Enterprise Approach in response to, or in anticipation of, changes in the nature of the mission or circumstances is supported by both logic and experimental results. One important implication of these findings is that enterprises need to invest in a capability to monitor their location in the Approach Space (if a collective, then for each entity as well as the collective).

9

ENTERPRISE OF THE FUTURE

The four megatrends identified in this book are having a profound impact on individual organizations and the complex endeavors they undertake. These trends pose unavoidable challenges for organizations acting alone or in concert with others; challenges that, if not adequately addressed, could result in catastrophic failures. As a result, militaries, civilian government agencies, businesses, nongovernmental organizations, and private voluntary organizations faced with "Big Problems" will increasingly find it more difficult to remain competitive and to succeed in the endeavors that they undertake unless they reinvent themselves. Fundamental changes to self and entity capabilities are necessary to develop the agility needed to survive in a complex and dynamic world. Those that are unwilling or unable to make the necessary changes risk being marginalized or disappearing.

A major thesis of this book is that the Enterprise of the Future—a military establishment or a complex coalition involving military and civilian partners, or some other entity—will be shaped by the actions it takes to meet these challenges. The Enterprise of the Future will likely need to be far more agile in order to survive and prosper. Developing overall enterprise agility begins with developing C2 Agility. This is because adopting an appropriate approach to Command and Control (C2) enables an enterprise to focus and bring to bear the resources at its disposal. It must do so first in an effort to understand the evolving situation, and then to orchestrate an effective, efficient, and timely response. Such a response must avoid, preempt, or mitigate adverse consequences, and seize opportunities to increase effectiveness and efficiency or reduce risk.

Fortunately, the four megatrends not only create existential challenges for organizations but also provide opportunities that they can seize to meet those challenges. Opportunities are created by the continuing maturation and proliferation of a global Robustly Networked

Environment or infostructure, and the set of sensor, information processing, and communications capabilities that have resulted in Ubiquitous Data. These capabilities and the economics associated with them have combined to expand greatly the universe of feasible approaches for C2, management, and governance available to organizations.

The Enterprises of the Future will need to take full advantage of the range of approach options available to them—that is, they must achieve C2 Agility.

9.1 Agile C2 for an Agile Enterprise

To acquire C2 Agility, enterprises will need to

- Develop a range of enterprise (C2/management/governance) approach options, particularly ones that are network enabled and can encompass a range of decentralized behavior as appropriate
- Understand which approach option is appropriate for the circumstances at hand
- Be able to shift from one Enterprise Approach to another in a timely and efficient manner
- Ensure access to desired regions of the Approach Space to avoid being denied the ability to operate effectively and efficiently

In order for an enterprise to be able to accomplish the above, it will need to develop, deploy, and employ appropriate approach options, organization policies and processes, personnel selection and promotion criteria, systems, education and training objectives, and programs.

9.2 Developing Enterprise Approach Options

A major driver of enterprise agility is the ability to adopt network-enabled approaches that involve broader, less centralized allocations of decision rights as appropriate. In Chapter 6, we discussed mission command, a command philosophy and family of approaches involving the delegation of decision rights to the lowest levels consistent with mission objectives, and the capabilities and resources of subordinates. The extent to which decision rights can be delegated with successful outcomes is constrained by the enterprise's ability to develop shared awareness and understanding. High levels of shared awareness are generally necessary to engender trust and ensure that subordinates will behave in an appropriate fashion. While Nelson at Trafalgar (see Chapter 1) demonstrated that sufficient levels of shared awareness, trust, and confidence can be achieved without today's advanced technical means, this naval engagement was not as large, complex, and dynamic as many of the missions undertaken in the 21st century. Given the multidimensionality of many of today's missions and the heterogeneous collection of entities required to bring appropriate information and resources to bear in a timely manner, developing the necessary levels of shared awareness requires widespread, cross-boundary information sharing and collaboration. A Robustly Networked Environment accompanied by Ubiquitous Data can provide the connectivity and the information to support the creation of the shared awareness needed to support a broad allocation of decision rights. At the same time, as we have seen in Chapters 4 and 5, these megatrends also empower adversaries and create the hazard of information overload, especially if the adversary entity is smaller, nimbler, and more dispersed, and has fewer data requirements of its own.

> The Enterprise of the Future will be able to adopt a range of network-enabled approaches, including approaches where decision rights are broadly distributed.

9.3 Understanding Which Approach Is Appropriate

Being able to adopt a variety of approaches to C2 has little value unless one chooses the right one. There are two basic selection strategies that can be employed in deciding which among the available C2 Approaches one should initially adopt.

The first strategy is to choose the most network-enabled (least-centralized) approach that the enterprise is capable of adopting. Evidence suggests that the more network enabled an approach is, the more agile it is (see Chapter 8). Therefore, the most network-enabled approach is likely to be successful in more of the Endeavor Space than other available approaches. This does not mean that this approach *will* be successful, given the mission and circumstance. It only suggests that the most network-enabled approach represents a best bet if there is no information that would suggest another choice.

The second strategy is to determine, to the best of one's analytic abilities with the available information, what the most appropriate choice would be, and adopt it. The approach decision calculus should take into consideration not only the current situation but how it could evolve and its implications for the ability to maintain the approach's location in the Approach Space. Thus, for certain routine operations, an enterprise may successfully choose some variation of a classic hierarchy, albeit with more connectivity and better information flows. For military operations, particularly in more chaotic environments, some type of mission command may be a good choice. For some other operations, such as attempting to counter a nimble, dispersed, and decentralized adversary, an enterprise may choose a more network-enabled approach.

> Thus, the Enterprise of the Future will be able to choose an appropriate C2 Approach from a variety of available options.

9.4 Shifting from One Approach to Another

Changes in circumstances are virtually certain to occur, and changes in the nature of the mission frequently occur. Enterprises need the capability to monitor the evolving situation and anticipate and detect changes that warrant reviewing the appropriateness of the current approach. They also need the capability to maneuver in the Approach Space in response.

Today's organizations tend to focus on monitoring their competitive space—that is, environmental conditions and adversary actions or likely plans. The Enterprise of the Future will need to do a much better job of also monitoring itself and its approach. Also, since many endeavors involve collaboration with other enterprises, one not only needs to monitor oneself but also how one is interacting with other entities. Thus, the Enterprise of the Future will be able to, in near real time, *collect and make sense* of the large number of information accesses and interactions that are taking place so that it knows where it, its partners, and the collective as a whole are operating in the Approach Space.

Being aware of where one is located in the Approach Space is the first step to being able to maneuver in this space. An enterprise also needs to know where it needs to be in the space, and how to get there.

That is, it needs to know what behaviors must be encouraged and facilitated and which ones must be discouraged. Such changes may, in addition to changes in the way decision rights are allocated, require different policies, processes, permissions, priorities, and incentives. They may require changes in the perceptions and attitudes of individuals, and challenge the very core of the enterprise's normal culture.

> The Enterprise of the Future will be able to maneuver appropriately in the C2 Approach Space.

9.5 Ensure Access to the Desired Regions of the Approach Space

Having selected a region in the Approach Space that is appropriate for a given undertaking and having maneuvered to this region if necessary, the Enterprise of the Future needs to ensure that it has and continues to have access to it. That is, the enterprise needs to prevent disruptions in its information and communications infrastructure from denying it the ability to locate where it wants. Having located there, it must prevent changes in circumstances from moving the C2 Approach to a location that is not appropriate for the mission.

Ironically, today's organizations are often voluntarily denying themselves significant portions of the Approach Space by their current policies and practices. Because of this self-inflicted wound, one of the most significant changes that will need to occur involves the way organizations view information "ownership" and "security." Traditionally, those who collect the information "own" the information. They determine if and with whom they will share it. The basis for sharing has traditionally been "need to know." For a brief period, the U.S. Department of Defense moved toward notions of collective ownership and a need to share. However, increased attacks on infostructure and information (or rather increased awareness of the problem) have returned the U.S. Department of Defense and many of its partner organizations to a defensive information-sharing policy stance rather than the more rational risk management approach that was beginning to take root.

There is a fine line between the legitimate need to maintain security and restricting information so much that one is essentially perpetrating a *denial-of-service attack on oneself.* The Enterprise of the

Future will need to develop information-sharing policies and practices supported by technology that promote and enable agile information dissemination. Without this, organizations will, in effect, deny themselves the ability to work in regions of the Approach Space that are the most appropriate for Big Problems.

> The Enterprise of the Future will be able to protect its position in the Approach Space. It will recognize when it is being relocated and reposition itself appropriately.

9.6 Beyond C2 Agility

While Agile C2 helps ensure that an enterprise can bring available information and material resources to bear in a timely manner, an enterprise also needs to have the human and technical resources that can provide it with the high level of shared understanding that underpins effective action and creates a competitive advantage. This begins with individuals who possess the domain knowledge, expertise, and experience necessary to anticipate and accomplish the tasks that are required. Increasingly, enterprises need to be able to take advantage of Ubiquitous Data, not only to enable them to manifest C2 Agility, but to make sense of situations, formulate plans, synchronize actions, and persistently monitor themselves, their adversaries, and the environment, making adjustments to both the C2 Approach and to operational plans, as required.

> This means that the Enterprise of the Future will need to be able to climb the "data to understanding" ladder, protect its infostructure, and develop and maintain appropriate levels of trust in the social and information domains.

9.7 From Data to Shared Understanding

9.7.1 Data Collection

The Robustly Networked Environment mediates an increasing proportion of the interactions we have, both economic and social. This is because we are connected, even when on the move. We use devices that are increasingly able to sense their environment, process and present information, and stay connected. As a result, sensors and

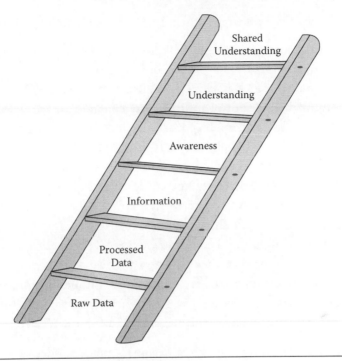

Figure 9.1   From data to shared understanding.

collectors are omnipresent. In the process, a huge amount of raw data is captured and made available to all who desire it, whenever they want it and wherever they are located.

Raw data is potentially useful, but only if an enterprise can move data up the rungs of the ladder to shared understanding (Figure 9.1). Enterprises have always had to contend with irrelevant, incorrect, and outdated data. But with the advent of Ubiquitous Data, enormous quantities of such noise gets swept up in the frenzied quest for needed information. A necessary task for the Enterprise of the Future will be to facilitate the discovery of potentially useful information, while at the same time efficiently reducing or eliminating noise.

The Enterprise of the Future will understand that it needs not only real-time information about others and the environment but also information about itself and its behaviors. It will, therefore, need to instrument itself and its systems.

The Enterprise of the Future will be able to collect and discover relevant data about its environment, its adversaries, its partners, and itself.

9.7.2 *Data to Understanding*

We have noticed that the voices raised in alarm at the specter of "information overload" have been out-shouted of late by those proclaiming an era of "Big Data."[*] Undeniably, there have been huge increases in the amount of information available and arguably some of it is more detailed, more persistent, and more accurate than has existed previously. The question is whether or not enterprises now or in the near future will have the tools they need to discover what they need to know in all this data and to do so in a timely, cost-effective fashion. This question is reflected in a white paper, "Unleashing the Potential of Big Data,"[†] that concludes that harnessing Big Data can, among other things, accelerate knowledge development and enhance productivity. The issue is if and when this capability moves from a promise to a practical capability. There are a number of problems that must be overcome, including the possibility of false correlations, and technical issues associated with database management.[‡]

> Data is a valuable resource, not a free good, and the Enterprise of the Future, shaped by competitive pressures, will possess a capability to selectively exploit Ubiquitous Data.

9.7.3 *Understanding to Shared Understanding*

Understanding enables individuals to accomplish tasks and do so more effectively and efficiently. Shared understanding unleashes enterprise agility. Collaboration and information sharing offer an opportunity to improve information quality and thus improve the quality of understanding. The more people who correctly understand what is required in a given situation, what options are available, and how to get something done, the greater the chance that the enterprise will detect changes in circumstances and respond appropriately in a timely manner.

> The Enterprise of the Future will create, and place a high value on, shared understanding.

---

[*] See Chapter 5.

[†] IBM et al. (2013).

[‡] Chapter 5; Arbesman (2013); Orenstein and Vassiliou (2014).

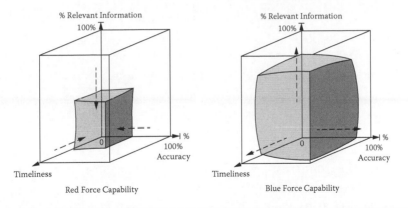

Figure 9.2 Dynamics of information superiority. (From Alberts et al., 1999. With permission.)

9.7.4 Assured Infostructure

Enterprises that cannot or will not assure their information and communications infrastructure will find themselves at a competitive disadvantage. The self-inflicted wounds they will suffer range from denying themselves an opportunity to operate in regions of the Approach Space that are appropriate to denying themselves information superiority.

Information superiority, a term that was introduced into the military vernacular in the 1990s,[*] is defined as "the ability to collect, process and disseminate an uninterrupted flow of information while exploiting and/or denying an adversary's ability to do the same."[†] Figure 9.2[‡] illustrates the dynamics of information superiority by comparing the information positions of "Red" (adversary) and "Blue," and the forces at work (illustrated by arrows) that create a competitive advantage in the information domain by increasing the relevance, timeliness, and accuracy of information. Not adequately protecting one's infostructure can result in the implosion of one's information position.

In an interconnected world, information vulnerabilities can be contagious, so enterprises that cannot assure their infostructure may not find many willing partners. Therefore, they will have limited opportunities to exchange information and interact with others over the

---

[*] For example, Alberts et al. (1999).

[†] USJCS (2012).

[‡] Alberts et al. (1999), Figure 7, p. 56.

network. Not only may an enterprise that does not assure its infostructure be relegated to an inferior information position, it will likely be unable to fully participate in coalitions and collectives, thus further limiting its ability to influence and react to events.

9.8 Trust

Trust, in a variety of forms, is essential for the Enterprise of the Future. A lack of trust creates impediments to the behaviors that are necessary to delegate decision rights, enhance one's information position, develop shared understanding, and realize operational synergies. We saw in Chapter 6 that trust in the training, capability, and motivations of subordinates is essential to mission command.

In the domain of information sharing, a lack of trust can freeze information in place, while appropriate trust assessments can move the right information along to the right places. Recalling Megatrends 2 and 3, the Robustly Networked Environment and Ubiquitous Data, extra vigilance is warranted. The importance of placing appropriate trust in the information we access and the individuals we share it with cannot be overemphasized in light of continuing efforts by newly capable adversaries to deny, degrade, compromise, or corrupt this asset.

The Enterprise of the Future will need to be aware of and monitor trust perceptions and work to ensure that individuals, information sources, and systems are deserving of trust and are appropriately trusted.

> The Enterprise of the Future will establish, value, and protect trusted relationships.

9.9 Key Questions for the Enterprise of the Future

The Enterprise of the Future will have, *of necessity*, achieved the ability to successfully tackle Big Problems, problems that are characterized by their high complexity and dynamics and the uncertainty and risk that accompanies them. Today's organizations need to know the answers to two questions to position themselves. The first is "what are the key attributes that my organization must have to succeed in the future?" The second is "what is different about the organizations that have developed these attributes and my organization as it is today?"

9.10 Key Attributes of the Enterprise of the Future

Based on the above discussion, we can identify the following key attributes of the successful Enterprise of the Future. It will be as follows:

- Agile and prize agility
- Robustly connected and task clustered
- Information savvy
- Willing and able to adopt a variety of network-enabled approaches to C2

9.10.1 Agile and Prizing Agility

The Enterprises of the Future will consist of those that have learned how to survive and prosper in a complex and dynamic world. Enterprises that have decided to bet their futures on being able to accomplish only the missions they deemed to be most likely, and ignoring others, will almost certainly be surprised and relatively unprepared to meet the challenges that they actually face in such a world. In a world that is shaped by competitive forces, these enterprises will either learn to value agility or be marginalized.

9.10.2 Robustly Connected—Task Clustered

Accomplishing complex and dynamic missions requires that enterprises have timely access to information, expertise in multiple domains (economic, political, military, etc.), and the ability to work with a set of heterogeneous partners. These all require that enterprises be not only robustly connected within themselves, but also robustly connected to a wide variety of information sources and to other enterprises. Robust connectivity between and among enterprises means that there must be multiple links between any two given enterprises so that the individuals within have direct access to each other and can dynamically form task groups as the need arises (see Figure 9.3).*

---

* Alberts et al. (2010), Figure 14.

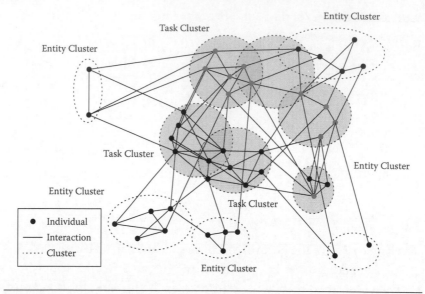

Figure 9.3   Robustly connected and task clustered. (From Alberts et al., 2010. With permission.)

9.10.3 Information Savvy

The ability to extract value from information will continue to represent a significant competitive advantage. Domain expertise is about mastering a given subject. Being information savvy is about an enterprise mastering the ability to collect appropriate information and then enabling those who need it to access it, when they need it, and in usable form.

9.10.4 Willing and Able to Adopt a Variety of Network-Enabled Approaches to C2

While being robustly connected and information savvy are two critical capabilities an enterprise needs to improve its agility, they are not sufficient to ensure enterprise agility. Being robustly connected and information savvy makes it possible to develop and maintain shared awareness, but to exhibit agility enterprises need to be able to bring assets to bear on problems and missions. That is, enterprises must be able to take appropriate, synchronized action in a timely manner, whatever the mission requirements or the circumstances. C2 Agility is related to the degree to which the C2 Approach is network enabled.

Thus, enterprise agility is a function of the willingness and ability of an enterprise to employ approaches to C2 that allocate decision rights broadly, and show related decentralized behavior in interaction patterns and information distribution supporting the broad allocation of decision rights.

> The Enterprise of the Future will prize agility, be robustly connected, task clustered, and information savvy, and be willing and able to adopt a variety of network-enabled approaches to C2.

9.11 Differences between Current Enterprises and the Enterprises of the Future

The Enterprises of the Future will differ significantly from many of today's organizations in the following ways:

- View of self
- Risk management
- Lack of inertia
- Infostructure capabilities

9.11.1 View of Self

While today's organizations tend to think of *themselves* as "the enterprise," the Enterprise of the Future will collaborate frequently with other enterprises and will learn to think of itself also as an integral part of the resulting collective. This broader perspective will permeate throughout whatever collective entity is constructed from the enterprise and its partners. While individuals will still consider themselves members of a specific organization, they will also identify with the collective. As a result, a level of trust will exist between and among collaborating entities and the individuals who belong to these entities, sufficient to facilitate the necessary flows of information and instances of collaboration required for the tasks at hand. This view will result in the development of processes and systems that can operate across entity boundaries. Human resource policies in each of the collaborating organizations may need to be adjusted so that individuals are properly incentivized to act in accordance with collective purposes. For example, if financial rewards and promotions are based exclusively on serving the needs of a component collaborating enterprise and do not reflect collective goals, the collective enterprise is more likely to fail.

9.11.2 Risk Management

The Enterprise of the Future will choose to manage rather than avoid risk. The Enterprise of the Future will understand that agility enhances the ability of individuals, organizations, and systems to manage a variety of risks. Hence, it will seek to continually improve its agility, rather than concentrating only on its effectiveness or efficiency.

9.11.3 Lack of Inertia

Inertia is the enemy of agility. Today's organizations exhibit high levels of resistance to change that prevent them from being more agile. In the area of C2, the impediments to Agile C2 include a culture of micromanagement, a go-it-alone attitude, a lack of trust, and a resultant unwillingness to share information and collaborate. These impediments deny the organizations of today access to a portion of the C2 Approach Space. As a result their C2 Agility is limited. The Enterprise of the Future will have removed or reduced these impediments, and having done so unlocked its agility potential.

9.11.4 Infostructure Capabilities

Many of today's organizations consider their infostructure to be a cost to be minimized rather than an investment that can yield large returns. They may focus on reducing the costs of the equipment that they own or lease, and the services they contract for. Such a focus on cost reduction may lead an enterprise to underestimate potential returns from the infostructure that can include efficiencies in operations, increases in measures of effectiveness, and enhanced agility. Many of today's organizations do not have adequate means to protect and assure their infostructure, and they lack the instrumentation and visualization capabilities to see how it is performing in real time.

The Enterprise of the Future will look beyond its own infostructure and consider how it can gain advantage from the global infostructure. In evaluating investments in equipment and services, the Enterprise of the Future will consider not only costs, but also the added functionality and agility that its own infostructure, and the interaction of that infostructure with the global one, can provide. Protecting the infostructure and assuring the information and services it delivers will be a priority.

9.12 The Way Ahead

The premise of this book is that there are four megatrends that will collectively shape the Enterprises of the Future. In the previous section, we took a logical leap to the future and returned with a vision of the values and capabilities possessed by the Enterprises of the Future. We also noted a number of key differences between the Enterprises of the Future and today's organizations. This information provides current organizations with the end-points of the journey each of them needs to take. That is, they know where they must go to remain successful and they know their point of departure (their current state).

Despite this knowledge, making this journey will still be extremely challenging. Today's organizations will need to change their sense of self; their values, incentives, and rewards; their management approaches; their communications, information collection, and analysis systems; their processes; and both individual and collective behaviors. These changes will affect virtually everyone.

The first, and perhaps the most important, step will be the development of an enterprise vision and its socialization, gaining widespread acceptance for it. There are thousands of decisions made and actions taken by members of an enterprise every day, and these decisions and actions actually determine whether or not an enterprise is successful. The purpose of an enterprise vision is to inform and shape these decisions. The development of such a vision must not be an empty exercise carried out by bureaucratic "task forces" producing briefing charts but no real understanding or acceptance. It must be real.

A sincere and accepted vision, while important, is still insufficient to effect the changes that will be required of today's organizations. The vision must be supported by appropriate attitudes, education, training, systems, and resources.

Organizations will have to take a hard look at their investment priorities. For example, in the face of budget constraints, should they continue to cut funds going to education and training and related professional activities as they have a tendency to do? Do investments in maintaining an arbitrary "end strength" rather than investing in a more agile force still make sense? Readers will be able to think of many questions like these if an enterprise commits to a future where agility is prized. Do investments in more platforms make sense in the absence of the investment needed to network them? Do investments in collection make sense in the absence of being able to protect and analyze the information collected?

Because the changes required are at every level, in each corner, and affect everyone, this transformation cannot be accomplished in a timely manner by a traditional approach to C2 or management. Change of this magnitude requires self-motivated, task-organized, self-synchronization informed and shaped by shared intent and understanding and continuous feedback. The widespread acceptance and internalization of the enterprise vision creates the necessary shared intent and also contributes to shared understanding. Appropriately instrumented systems and environments can provide the necessary feedback. A robustly connected infostructure enables all of the above.

First and last, however, the willingness to delegate decision rights turns the potential inherent in well-educated and trained individuals into manifestations of the agility necessary to succeed in a complex and dynamic world. Lofty words and sham delegations will only breed cynicism and will not work. Decision rights must be truly delegated, and that delegation must be *defended*. One of the reasons for the successful development of mission command in the Prussian and German armies of the 19th and early 20th centuries was that senior commanders learned to stand behind the principle of delegation to subordinates, even if it sometimes yielded unsuccessful results. We saw in Chapter 6 that von Moltke, in analyzing the performance of the defeated Austrians at the Battle of Königgrätz, defended the independent action that two Austrian generals had taken in defiance of their superior commander, even though that action had helped the Prussians win. He believed their actions were justified under the circumstances, and he believed in the principle of subordinate initiative. He understood that a war is too complex and chaotic a phenomenon to be centrally micromanaged.

Glossary of Acronyms

2G: Second-Generation Cellular Telephone Technology
3G: Third-Generation Cellular Telephone Technology
4G: Fourth-Generation Cellular Telephone Technology
abELICIT: Agent-Based ELICIT
ACTS: Accuracy, Completeness, Timeliness, and Standards
ADR: Allocation of Decision Rights
ADS: Authoritative Data Source
ADSL: Army Data Services Layer
ADT: Army Data Transformation (United States)
ANCDS: Army Net-Centric Data Strategy
AO: Area of Operation
API: Application Programming Interface
ARGUS-IS: Autonomous Real-Time Ground Ubiquitous Surveillance Imaging System
C2: Command and Control
C2IEDM: Command and Control Information Exchange Data Model
C3: Command, Control, and Communications
C4: Command, Control, Communications, and Computers
C4ISR: Command, Control, Communications, Computers, Intelligence, Surveillance, and Reconnaissance
CDC: Centers for Disease Control and Prevention (United States)

CDMA: Code Division Multiple Access; a commercial cellular technology

CFA: Country Fire Authority (State of Victoria, Australia)

CIDNE: Combined Information Data Network Exchange

COP: Common Operating Picture

COTS: Commercial Off-the-Shelf, or Common Off-the-Shelf

CPOF: Command Post of the Future

DAMO: Department of the Army, Military Operations (United States)

DARPA: Defense Advanced Research Projects Agency (United States)

DHS: Department of Homeland Security (United States)

DISA: Defense Information Systems Agency (United States)

DISN: Defense Information Systems Network (United States)

DoD: Department of Defense (United States)

DoI: Distribution of Information

DOTMLPF: Doctrine, Organization, Training, Material, Leadership, and Facilities

DSE: Department of Sustainability and the Environment (State of Victoria, Australia)

EDGE: Enhanced Data Rates for GSM Evolution; a commercial cellular technology based on GSM

ELICIT: Experimental Laboratory for the Investigation of Collaboration, Information-Sharing, and Trust

EPLRS: Enhanced Position Location Reporting System

ETA: Euskadi Ta Askatasuna

ETL: Extract-Transform-Load

EU: European Union

EU-28: The 28 member nations of the European Union as it stood in 2014

EV-DO: Evolution-Data Optimized; an enhancement of the CDMA-2000 cellular technology

FAA: Federal Aviation Administration (United States)

FDNY: Fire Department of New York (New York City)

FEMA: Federal Emergency Management Agency (United States)

FPGA: Field Programmable Gate Array

FSCL: Fire Service Control Line

Gbps: Gigabits per second

GCCS: Global Command and Control System

GPRS: General Packet Radio Service; a commercial cellular technology based on GSM

GSM: Global System Mobile; a commercial cellular technology

HF: High Frequency

HQ: Headquarters

HSDPA: High-Speed Downlink Packet Access; a commercial cellular technology based on UMTS

HSPA+: High-Speed Packet Access +; a commercial cellular technology based on UMTS

IAGCF: Intelligent Agents Computer Generated Forces

IC: Intelligence Community

ICT: Information and Communications Technology

IDF: Israeli Defense Forces

IMT: International Mobile Telephony

IP: Internet Protocol

IPM: Information Production Maps

IRA: Irish Republican Army

IS95A: Interim Standard 95A; based on Code Division Multiple Access, a commercial cellular technology

IS95B: Interim Standard 95B; based on Code Division Multiple Access, a commercial cellular technology

ISO: International Standardization Organization

ISR: Intelligence, Surveillance, and Reconnaissance

IT: Information Technology

ITU: International Telecommunication Union

JC3IEDM: Joint Consultation, Command, and Control Information Exchange Data Model

JCA: Joint Capability Area (United States)

JSON: JavaScript Object Notation

JTRS: Joint Tactical Radio System

LINPACK: LINear equations software PACKage

LTE: Long-Term Evolution; a commercial cellular technology

Mbps: Megabits per second

Mega FLOPS: Mega Floating Point Operations per Second

MFLOPS: Million Floating Point Operations per Second, or Mega Floating Point Operations per Second

MIDS: Multifunctional Information Distribution System

MIDS-LVT: MIDS Low-Volume Terminal

N2M2C2: NATO Network-Enabled Capability C2 Maturity Model (see SAS 065)

NATO: North Atlantic Treaty Organization

NCDS: Net-Centric Data Strategy (U.S. Department of Defense)

NCW: Net-Centric Warfare

NEC: Network-Enabled Capability

NGO: Nongovernmental Organization

NIEM: National Information Exchange Model

NLP: Natural Language Processing

NNEC: NATO Network-Enabled Capability

NORAD: North American Aerospace Defense Command (United States)

NoSQL: Not SQL, or Not Only SQL

NSA: National Security Agency (United States)

NTDR: Near-Term Digital Radio

NYPD: New York Police Department (New York City)

OECD: Organization for Economic Cooperation and Development

OGD: Other Government Department

OODA: Observe, Orient, Decide, Act

OS: Operating System

PANOPEA: Piracy Asymmetric Naval Operation Patterns Modeling for Education and Analysis

PAPD: Port Authority Police Department (Port Authority of New York and New Jersey)

PED: Processing, Exploitation, and Dissemination

PoI: Patterns of Interaction

R99: Release 99; a commercial cellular technology, based on UMTS

SAS: System and Analysis Studies (NATO)

SAS 050: NATO System and Analysis Studies group 050 Exploring New Command and Control Concepts and Capabilities

SAS 065: NATO System and Analysis Studies group 065: Network-Enabled Capability (NNEC) C2 Maturity Model

SAS 085: NATO System and Analysis Studies group 085: C2 Agility

SINCGARS: Single Channel Ground and Airborne Radio System

SINCGARS SIP: SINCGARS System Improvement Program

SNR: Signal-to-Noise Ratio

SQL: Structured Query Language

SRW: Soldier Radio Waveform (JTRS)
STO: Science and Technology Organization (NATO)
SWAP: Size, Weight, and Power
TDQM: Total Data Quality Management
TIGR: Tactical Ground Reporting System
UAV: Unmanned Aerial Vehicle
UCore: Universal Core
UGV: Unmanned Ground Vehicle
UHF: Ultrahigh Frequency
UMTS: Universal Mobile Telecommunications System; a commercial cellular technology, based on GSM
UN: United Nations
USMC: U.S. Marine Corps
VHF: Very High Frequency
VVS: Voyenno-Vozdushnyye Sily (Russian Air Force)
WIMAX: Worldwide Interoperability for Microwave Access; a commercial cellular and wireless networking standard
WISE: War Game Infrastructure and Simulation Environment
WNW: Wideband Networking Waveform (JTRS)
XML: Extensible Markup Language

Appendix A

In Chapter 1, we defined Command and Control (C2) as follows

> "Command and Control" (C2) denotes the set of organizational and technical attributes and processes by which an enterprise marshals and employs human, physical, and information resources to solve problems and accomplish missions.

There are many other definitions of C2 in the literature and in common use. None are universally accepted. Most can be subsumed by the definition above.

A.1 More Restrictive Definition in the Business Community

In much of the business and organizational literature, "Command and Control" denotes a particular style of top-down management. For example, the *Bloomsbury Business Dictionary* states that a "Command and Control Approach" denotes a

> style of leadership that uses standards, procedures, and output statistics to regulate the organization. A command and control approach to leadership is authoritative in nature and uses a top-down approach, which

fits well in bureaucratic organizations in which privilege and power are vested in senior management.*

In this book, we consider this a subset of the overall definition, denoting a particular hierarchical C2 Approach.

A.2 Official Military Definitions

The U.S. Department of Defense's *Dictionary of Military and Associated Terms* offers the following definitions:

> Command and Control—The exercise of authority and direction by a properly designated commander over assigned and attached forces in the accomplishment of the mission. Also called C2.†
>
> Command and Control System—The facilities, equipment, communications, procedures, and personnel essential to a commander for planning, directing, and controlling operations of assigned and attached forces pursuant to the missions assigned.‡

These are consistent with our more general definition. The North Atlantic Treaty Organization (NATO) defines a "Command and Control System" as "An assembly of equipment, methods and procedures and, if necessary, personnel, that enables commanders and their staffs to exercise command and control,"§ but does not define "Command and Control," at least not as a phrase. NATO does separately define "Command" and "Control" as words. It defines the former as "The authority vested in an individual of the armed forces for the direction, coordination, and control of military forces," and the latter, rather unhelpfully, as "That authority exercised by a commander over part of the activities of subordinate organizations, or other organizations not normally under his command, which encompasses the responsibility for implementing orders or directives. All or part of this authority may be transferred or delegated." ¶

---

* Bloomsbury (2007), p. 1651.
† USDoD (2014), p. 45.
‡ Ibid.
§ NATO (2008), p. 2-C-9.
¶ NATO (2008), pp. 2-C-9 and 2-C-14.

A.3 Pigeau and McCann

Pigeau and McCann (2002) conducted a survey of definitions of C2 and found the official ones inconsistent and sometimes circular and redundant. They proposed instead a new definition of *Command* as the "creative expression of the human will necessary to accomplish the mission," and *Control* as "those structures and processes devised by command to enable it and to manage risk."* These definitions are consistent with the one we have adopted.

A.4 Capability-Based Definition

The office of the U.S. Department of Defense's Joint Chiefs of Staff has classified military capabilities into nine "Joint Capability Areas" (JCAs) for the purposes of strategy development, force development, operational planning, technological portfolio management, and investment planning.† One of the JCAs is *Command and Control*. The main capabilities in the C2 JCA are *organizing*, *understanding*, *planning*, *deciding*, *directing*, and *monitoring*. These, and several finer-grained capabilities, constitute a capability or activity-based definition of C2. More detail is shown in Figure A.1.

Figure A.1 also shows the three other closely related JCAs of *Net-Centric*, *Battlespace Awareness*, and *Building Partnerships*.‡ C2 is very closely linked to these other capability areas and highly dependent on some of them for input. In particular, it is inextricably linked to the *Net-Centric* JCA. Readers may be familiar with some of the expanded acronyms that have cropped up over the years, such as "Command, Control, and Communications" (C3), which recognizes the crucial importance and inseparability of communications from C2. C3 is also often expanded to C4, "Command, Control, Communications, and Computers," recognizing the central place of computing and information technologies in modern operations. Finally, a common acronym in regular use is C4ISR, or "Command, Control, Communications, Computers, Intelligence, Surveillance,

---

* Pigeau and McCann (2002), p. 56.
† USJS J-7 (2008, 2014).
‡ The five other JCAs are Force Support, Force Application, Logistics, Protection, and Corporate Management and Support.

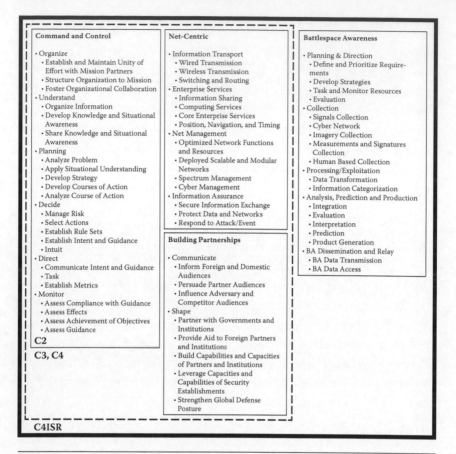

Figure A.1 C2, C3, C4, and C4ISR in the context of the U.S. Department of Defense's Joint Capability Areas. (Adapted from Vassiliou, 2010.)

and Reconnaissance," which recognizes C2 and its inputs as an organic whole. As seen in Figure A.1, a combination of the *C2*, *Net-Centric*, *Building Partnerships*, and *Battlespace Awareness* JCAs largely covers the generic capabilities necessary for C4ISR.

C4ISR is central to modern warfare, and C2 is central to C4ISR. C4ISR also represents some of the most technologically intensive military capabilities. A keyword analysis of the U.S. Department of Defense's research and development portfolio for fiscal years 2010 through 2012 revealed that technical areas broadly interpretable as supporting C4ISR accounted for around 40% of total spending.[*]

---

[*] Kramer et al. (2012).

Appendix B

Data Sizes

TERM	BINARY SYSTEM[a,b]	DECIMAL SYSTEM[c]
Kilobyte	$2^{10} = 1.0240 \times 10^3$ bytes	10^3 bytes
Megabyte	$2^{20} = 1.0486 \times 10^6$ bytes	10^6 bytes
Gigabyte	$2^{30} = 1.0737 \times 10^9$ bytes	10^9 bytes
Terabyte	$2^{40} = 1.0995 \times 10^{12}$ bytes	10^{12} bytes
Petabyte	$2^{50} = 1.1259 \times 10^{12}$ bytes	10^{15} bytes
Exabyte	$2^{60} = 1.1529 \times 10^{18}$ bytes	10^{18} bytes
Zettabyte	$2^{70} = 1.1806 \times 10^{21}$ bytes	10^{21} bytes
Yottabyte	$2^{80} = 1.2089 \times 10^{24}$ bytes	10^{24} bytes

Note: One byte = 8 bits.

[a] The binary system is the "proper" one, used in computer science.

[b] Right-hand side of equation given to four decimal places.

[c] The decimal system is widely used by manufacturers of storage media, and by the public at large.

References

Aboutnelson.co.uk (2014). Nelson's Iconic "England Expects" Signal. http://www.aboutnelson.co.uk/england%20expcts.htm (accessed August 8, 2014).

AeroVironment (2014). Puma AE Datasheet. Monrovia, California: AeroVironment Corporation. http://www.avinc.com/downloads/DS_Puma_Online_10112013.pdf (accessed August 8, 2014).

Agre, Jonathan, M. S. Vassiliou, and Corinne Kramer (2011). Science and Technology Issues Relating to Data Quality in C2 Systems, *Proc. 16th International Command and Control Research and Technology Symposium (ICCRTS)*, Québec City, Québec, Canada, June 21–23. Washington, DC: CCRP Press. Paper 031, pp. 1–23. http://www.dodccrp.org/events/16th_iccrts_2011/papers/031.pdf (accessed August 8, 2014).

Agre, Jonathan, Karen Gordon, and Marius Vassiliou (2013). Commercial Technology at the Tactical Edge. *Proc. 18th International Command and Control Research and Technology Symposium (ICCRTS)*.

Albers, Michael J. (2011). Human-Information Interactions with Complex Software, First International Conference, DUXU 2011, Held as Part of HCI International 2011, Orlando, FL, July 9–14, *Proceedings*, Part II, Ed. Aaron Marcus (pp. 245–254). Berlin–Heidelberg: Springer.

Alberts, David S., John J. Garstka, and Frederick P. Stein (1999). *Network Centric Warfare: Developing and Leveraging Information Superiority*. 2nd ed. reprinted 2000. Washington, DC: U.S. Department of Defense, Command and Control Research Program (CCRP Press).

Alberts, David S., John J. Garstka, Richard E. Hayes, and David A. Signori (2001). *Understanding Information Age Warfare*. Washington, DC: U.S. Department of Defense, Command and Control Research Program (CCRP Press).

Alberts, David S., and Richard E. Hayes (2003). *Power to the Edge: Command and Control in the Information Age.* Washington, DC: U.S. Department of Defense, Command and Control Research Program (CCRP Press).

Alberts, David S., and Richard E. Hayes (2006). *Understanding Command and Control.* Washington, DC: U.S. Department of Defense, Command and Control Research Program (CCRP Press).

Alberts, David S., and Richard E. Hayes (2007). Planning: Complex Endeavors. Washington, DC: U.S. Department of Defense, Command and Control Research Program (CCRP Press).

Alberts, David S., and Mark E. Nissen (2009). Toward Harmonizing Command and Control with Organization and Management Theory. *International C2 Journal,* Vol. 3, No. 2, 1–59.

Alberts, David S., Reiner K. Huber, and James Moffat (2010). *NATO NEC C2 Maturity Model.* Washington, DC: U.S. Department of Defense, Command and Control Research Program (CCRP Press).

Alberts, David S. (2011). *The Agility Advantage.* Washington, DC: U.S. Department of Defense, Command and Control Research Program (CCRP Press).

Alberts, David S., and Marco Manso (2012). Operationalizing and Improving C2 Agility: Lessons from Experimentation. *Proc. 17th International Command and Control Research and Technology Symposium.* Fairfax, VA, June 19-21. Washington, DC: CCRP Press. Paper 086, pp. 1-32. http://www.dodccrp.org/events/17th_iccrts_2012/post_conference/papers/086.pdf (accessed August 8, 2014).

Alberts, David S., Marius Vassiliou, and Jonathan Agre (2012). C2 Information Quality: An Enterprise Systems Perspective. *Proc. IEEE Milcom 2012* (pp. 1–7). Hoboken, NJ: IEEE Press.

Alberts, David S., François Bernier, Kevin Chan, and Marco Manso (2013). C2 Approaches: Looking for the "Sweet Spot." *Proc. 18th International Command and Control Research and Technology Symposium,* Alexandria, VA, June 19–21. Washington, DC: CCRP Press. Paper 034, pp. 1–22. http://dodccrp.org/events/18th_iccrts_2013/post_conference/papers/034.pdf (accessed August 8, 2014).

Andrews, Jeffrey, Sanjay Shakkottai, Robert Heath, Nihar Jindal, Martin Haenggi, Randy Berry, Dongning Guo, Michael Neely, Steven Weber, Syed Jafar, and Aylin Yener (2008). Rethinking Information Theory for Mobile Ad Hoc Networks. *IEEE Communications Magazine,* December 2008, 94–101.

Anno, Stephen E., and William E. Einspahr (1988). Command and Control and Communications Lessons Learned: Iranian Rescue, Falklands Conflict, Grenada Invasion, Libya Raid. Air War College Research Report No. AU-AWC-88-043. Maxwell Air Force Base, AL: Air War College.

Appirio (2013). The Emerging Social Intranet: The 2013 Appirio Employee Portal/Social Intranet Survey. San Francisco, CA: Appirio. http://www.appirio.com (accessed August 8, 2014).

Arbesman, Samuel (2013). 5 Myths about Big Data. *The Washington Post*, August 18, 2013.

AT&T (2012). AT&T to Develop Highly-Secure, Commercially-Available Mobile Devices for Military and Enterprise. *PR Newswire*. Oakton, VA, December 19, 2012. http://www.bloomberg.com/article/2012-12-19/aIAC_fSKajZ0.html (accessed August 8, 2014).

Au, Andrew (2011). Analysis of command and control networks on Black Saturday. *Australian Journal of Emergency Management*, Vol. 26, No. 3, 20–29.

Avery, Benjamin, Ross T. Smith, Wayne Piekarski, and Bruce H. Thomas (2010). *Engineering Mixed Reality Systems*. New York: Springer.

Azuma, Ronald (1997). A Survey of Augmented Reality. *Presence: Teleoperators and Virtual Environments*, Vol. 6, No. 4, 355–385.

Ball, Peter (2006). Transcription of Telegraphic Signals or Marine Vocabulary 1801 by Sir Home Popham. http://3decks.pbworks.com/f/Admiral%2520Home%2520Popham%2520Telegraph%2520signal%2520book%2520Final%2520edition.pdf (accessed August 8, 2014).

Ballou, D.P., R.Y. Wang, H. Pazer, and G.K. Tayi (1998). Modeling information manufacturing systems to determine information product quality. *Management Science*, Vol. 44, No. 4, 462–484.

Barham [Charles, Lord Barham] (1805). Instructions to Lord Nelson, letter dated September 5, 1805. In: Laughton, Sir John K. Ed. (1911). *Letters and papers of Charles, Lord Barham, Admiral of the Red Squadron, 1758–1813*. Vol. III, Publications of the Navy Records Society, 39. London: Navy Records Society.

Basehart, Harry W. (1970). Mescalero Apache Band Organization and Leadership. *Southwestern Journal of Anthropology*, Vol. 26, No. 1, 87–106.

BBC (2010). Saudi Arabia Begins Blackberry Ban, Users Say. British Broadcasting Corporation, August 6, 2010. http://www.bbc.co.uk/news/world-middle-east-10888954. Retrieved February 22, 2011.

Behringer, Reinhold, C. Tam, J. McGee, V. Sundareswaran, and M.S. Vassiliou (2000a). A Wearable Augmented Reality Testbed for Navigation and Control, Built Solely with Commercial Off-the-Shelf (COTS) Hardware. *Proc. IEEE/ACM International Symposium on Augmented Reality* (pp. 12–19). Los Alamitos, CA: IEEE Computer Society Press.

Behringer, Reinhold, C. Tam, J. McGee, V. Sundareswaran, and M.S. Vassiliou (2000b). Two Wearable Testbeds for Augmented Reality: itWARNS and WIMMIS. *Proc. 4th International Symposium on Wearable Computers* (pp. 189–190).

Bennett, Brian (2013). Quad-Core Smartphone Shootout. *CNET Reviews*. http://reviews.cnet.com/8301-6452_7-57558350/quad-core-smartphone-shootout/ (accessed August 8, 2014).

Berman, Jules K. (2013). *Principles of Big Data: Preparing, Sharing, and Analyzing Complex Information*. New York: Elsevier.

Berman, M.G., J. Jonides, and S. Kaplan (2008). The Cognitive Benefits of Interacting with Nature. *Psychological Science*, Vol. 19, No. 12, 1207–1212.

Bernier, F. (2012). Agility of C2 Approaches and Requisite Maturity in a Comprehensive Approach Context. *Proc. 17th International Command and Control Research and Technology Symposium,* Fairfax, VA, June 19-21. Washington, DC: CCRP Press. Paper 081, pp. 1–17. http://www.dodccrp.org/events/17th_iccrts_2012/post_conference/papers/081.pdf (accessed August 8, 2014).

Best, Jo (2014). IBM Watson: The Inside Story of How the Jeopardy-Winning Supercomputer Was Born, and What It Wants to Do Next. *TechRepublic.* http://www.techrepublic.com/article/ibm-watson-the-inside-story-of-how-the-jeopardy-winning-supercomputer-was-born-and-what-it-wants-to-do-next/# (accessed August 8, 2014).

Biddle, Stephen, and Jeffrey A. Friedman (2008). The 2006 Lebanon Campaign and the Future of Warfare: Implications for Army and Defense Policy. Carlisle, PA: U.S. Army War College, Strategic Studies Institute.

Blake, Anne M., and James L. Moseley (2011). Frederick Winslow Taylor: One Hundred Years of Managerial Insight. *International Journal of Management,* Vol. 28, No. 4, Part 2, 349.

Bloomsbury Business Library (2007). *Business and Management Dictionary.* London: A&C Black, p. 1651.

Bobic, Michael P., and William Eric Davis (2003). A Kind Word for Theory X: Or Why So Many Newfangled Management Techniques Quickly Fail. *Journal of Public Administration Research and Theory,* Vol. 13, no. 3, pp. 239–264.

Borne, Kirk (2010). Astroinformatics: Data-Oriented Astronomy Research and Education. *Journal of Earth Science Informatics,* Vol. 1, No. 3, 5–17.

Bowden, Mark (2006). The Desert One Debacle. *The Atlantic Monthly,* May 2006, pp. 62–77.

Boyd, John (1995). The Essence of Winning and Losing. Unpublished slides, June 28, 1995, preserved at http://www.danford.net/boyd/essence.htm (accessed August 8, 2014).

Brafman, Ori, and Rod A. Beckstrom (2006). *The Starfish and the Spider: The Unstoppable Power of Leaderless Organizations.* New York: Portfolio (Penguin Group).

Braun, Torsten, Andreas Kassler, Maria Kihl, Veselin Rakocevic, Vasilios Siris, and Geert Heijenk (2009). Multihop Wireless Networks. Chapter 5 in Y. Koucheryavy et al. (eds.), *Traffic and QoS Management in Wireless Multimedia Networks, Lecture Notes in Electrical Engineering,* 31, pp. 201–265. New York: Springer.

Bravata, D.M., and I. Olkin (2001). Simple Pooling versus Combining in Meta-Analysis. *Evaluation and the Health Professions,* Vol. 24, No. 2, 218–230.

Brogan, Jesse W. (2011). Exonerating Frederick Taylor: After 100 Years, Mythology Sometimes Overshadows a Master's Teachings. *Industrial Engineer,* Vol. 43, No. 11, 41–44.

Bruzzone, A.G., A. Tremori, and Y. Merkuryev (2011a). Asymmetric Marine Warfare: PANOPEA a Piracy Simulator for Investigating New C2 Solutions. *Proc. SCM MEMTS Conference* (p. 32). St. Petersburg, Russia.

Bruzzone, A.G., M. Massei, F. Madeo, F. Tarone, and M.M. Gunal (2011b). Simulating Marine Asymmetric Scenarios for Testing Different C2 Maturity Levels. *Proc. 16th International Command and Control Research and Technology Symposium.* Québec City, Québec, Canada, June 21–23. Washington, DC: CCRP Press. Paper 192, pp. 1–13. http://www.dodccrp.org/events/16th_iccrts_2011/papers/192.pdf (accessed August 8, 2014).

Bungay, Stephen (2005). The Road to Mission Command: The Genesis of a Command Philosophy. *British Army Review,* Summer, Vol. 137, 22–29.

Bungay, Stephen (2011). *The Art of Action: How Leaders Close the Gaps Between Plans, Actions and Results.* Boston: Nicholas Brealey.

Burgess, Alan, and Peter Fisher (2008). *A Framework for the Study of Command and Control Structures.* Edinburgh, South Australia: Commonwealth of Australia Defense Science and Technology Organization, Publication DSTO-TN-0826.

Canada DND (Department of National Defence) (1996). *Conduct of Land Operations—Operational Level Doctrine for the Canadian Army.* Publication B-GL-300-001/FP-000. Ottawa, Ontario: Queen's Printer.

Carter, Joseph A., Shiraz Maher, and Peter R. Neumann (2014). *#Greenbirds: Measuring Importance and Influence in Syrian Foreign Fighter Networks.* London, UK: International Centre for the Study of Radicalisation and Political Violence, Department of War Studies, King's College. http://icsr.info/wp-content/uploads/2014/04/ICSR-Report-Greenbirds-Measuring-Importance-and-Infleunce-in-Syrian-Foreign-Fighter-Networks.pdf (accessed August 8, 2014).

Cartwright, James E. (2006). Information Sharing Is a Strategic Imperative. *Crosstalk,* Vol. 19, No. 7, 7–9. http://www.dtic.mil/dtic/tr/fulltext/u2/a488324.pdf (accessed August 8, 2014).

CCRP (2010). ELICIT Overview. U.S. Department of Defense, Command and Control Research Program. http://www.dodccrp.org/parity_files/ELICIT%20Overview%20v2.0_12.13.10.pdf. Retrieved February 24, 2012.

Cebrowski, Arthur K., and John H. Gartska (1998). Network-Centric Warfare, Its Origins and Future. *Proc. U.S. Naval Institute,* Vol. 124, No. 1, 139.

Cebrowski, A.K. (2003). Network-centric warfare. An emerging military response to the information age. *Military Technology,* Vol. 27, No. 5, 16–18.

Ceruti, M. (2003). Data Management Challenges and Development for Military Information Systems. *IEEE Transactions on Knowledge and Data Engineering,* Vol. 15, No. 5, 1059–1068.

Chan, K., and S. Adali (2012). An Agent Based Model for Trust and Information Sharing in Networked Systems. *Proc. Cognitive Methods in Situation Awareness and Decision Support (CogSIMA)* (pp. 88–95). New Orleans, LA. Hoboken, NJ: IEEE Press.

Chan, K., J.H. Cho, and S. Adali (2012). Composite Trust Model for an Information Sharing Scenario. *Proc. Ubiquitous Intelligence and Computing and 9th International Conference on Autonomic and Trusted Computing (UIC/ATC)* (pp. 439–446). Fukuoka, Japan. Los Alamitos, CA: IEEE Computer Society Press.

Cisco (2012). *Cisco Visual Networking Index: Global Mobile Data Traffic Forecast Update, 2011–2016*. San Jose, CA: Cisco Systems. http://www.cisco.com/en/US/solutions/collateral/ns341/ns525/ns537/ns705/ns827/white_paper_c11-520862.pdf (accessed February 21, 2012).

Cisco (2013). *Cisco Visual Networking Index: Forecast and Methodology, 2012–2017*. May 29. San Jose, CA: Cisco Systems. http://www.cisco.com/c/en/us/solutions/collateral/service-provider/ip-ngn-ip-next-generation-network/white_paper_c11-481360.html.

Cisco (2014). *Cisco Visual Networking Index: Global Mobile Data Traffic Forecast Update, 2013–2018*. Cisco, February 5. San Jose, CA: Cisco Systems. http://www.cisco.com/c/en/us/solutions/collateral/service-provider/visual-networking-index-vni/white_paper_c11-520862.pdf.

Citino, Robert (2004). Beyond Fire and Movement: Command, Control and Information in the German Blitzkrieg. *Journal of Strategic Studies*, Vol. 27, No. 2, 324–344.

Cohen, Ariel, and Robert E. Hamilton (2011). *The Russian Military and the Georgia War: Lessons and Implications*. Carlisle, PA: Strategic Studies Institute, U.S. Army War College. http://www.strategicstudiesinstitute.army.mil/pdffiles/PUB1069.pdf (accessed August 8, 2014).

Cohen, S., S. Cohen-Boulakia, and S. Davidson (2006). Towards a Model of Provenance and User Views in Scientific Workflows. *Data Integration in the Life Sciences, Lecture Notes in Computer Science*, Vol. 4075, 264–279.

Cole, Ronald H. (1997). *Operation Urgent Fury: The Planning and Execution of Joint Operations in Grenada, October 12–November 2, 1983*. Washington, DC: Office of the Chairman of the Joint Chiefs of Staff.

Collings, Deirdre, and Rafal Rohozinski (2009). *Bullets and Blogs—New Media and the Warfighter*. Carlisle Barracks, PA: U.S. Army War College.

Commission on 22 July (2012). *Norwegian Government Investigation 2012:14, Report of the Commission on July 22: Preliminary English Version of Selected Chapters*. [Original: Norges Offentlige Utredninger 2012:14, Rapport fra 22.juli kommisjonen]. Oslo, Norway: Government of Norway, Commission on July 22.

Cordesman, Anthony (2006). *Preliminary "Lessons" of the Israeli–Hezbollah War*. Washington, DC: Center for Strategic and International Studies.

CRA (2012). *Challenges and Opportunities with Big Data*. Washington, DC: Computing Research Association. http://www.cra.org/ccc/files/docs/init/bigdatawhitepaper.pdf (accessed August 8, 2014).

Croome, Desmond F., and Alan Arthur Jackson (1993). *Rails through the Clay: A History of London's Tube Railways*. London: Capital Transport Publishing.

Curran, P.J., and A.M. Hussong (2009). Integrative Data Analysis: The Simultaneous Analysis of Multiple Data Sets. *Psychological Methods*, Vol. 14, No. 2, 81–100.

Davenport, Thomas H., and D.J. Patil (2012). Data Scientist: The Sexiest Job of the 21st Century. *Harvard Business Review*, Vol. 90, No. 10, 70–76.

Day, Michael (2008). Current and Emerging Scientific Data Curation Practices. Presentation from 4th Summer School on Preservation in Digital Libraries, Tirrenia, Italy, June 12. http://www.slideshare.net/michaelday/research-data (accessed August 11, 2014).

Dean, Jeffrey, and Sanjay Ghemawat (2004). MapReduce: Simplified Data Processing on Large Clusters. *Proc. 6th Symposium on Operating Systems Design and Implementation.* https://www.usenix.org/legacy/publications/library/proceedings/osdi04/tech/full_papers/dean/dean_html/ (accessed August 11, 2014).

Defense Industry Daily (2005). Four-Star Blogging at STRATCOM. March 28. http://www.defenseindustrydaily.com/fourstar-blogging-at-stratcom-0239/ (accessed August 11, 2014).

Dempsey, Martin (2012). Mission Command White Paper. Washington, DC: U.S. Department of Defense, Office of the Chairman of the Joint Chiefs of Staff. http://www.dtic.mil/doctrine/concepts/white_papers/cjcs_wp_missioncommand.pdf (accessed August 11, 2014).

Dennis, Steve (2012). Independent Inquiry into Norway Killings Reveals Significant Failings by Police. London: Desroches Group. http://www.desrochesgroup.com/investigation-academy-articles/department-of-serious-crime-articles/independent-inquiry-into-norway-killings-reveals-significant-failings-by-police/ (accessed August 11, 2014).

DISA (2012). Joint C2: *Situational Awareness and Intel Global Command and Control System–Joint (GCCS–J). Briefing v1.4, DISA Mission Partners Conference.* Fort Meade, MD: Defense Information Systems Agency.

Dixon, Joshua, Geoffrey G. Xie, and Frank Kragh (2010). Integrating Cellular Handset Capabilities with Military Wireless Communications. *Proc. 15th International Command and Control Research and Technology Symposium (ICCRTS),* Santa Monica, CA, June 22–24. Washington, DC: CCRP Press. Paper 049, pp. 1–9. http://www.dodccrp.org/events/15th_iccrts_2010/papers/049.pdf (accessed August 11, 2014).

Donahue, Amy K., and Robert V. Tuohy (2006). Lessons We Don't Learn: A Study of the Lessons of Disasters, Why We Repeat Them, and How We Can Learn Them. *Homeland Security Affairs,* Vol. II, No. 2, 1–28.

Dongarra, J., J. Bunch, C. Moler, and G.W. Stewart (1979). *LINPACK User's Guide.* Philadelphia, PA: Society for Industrial and Applied Mathematics.

Dongarra, Jack J., Piotr Luszczeky, and Antoine Petitetz (2001). The LINPACK Benchmark: Past, Present, and Future. http://www.netlib.org/utk/people/JackDongarra/PAPERS/hpl.pdf. Retrieved February 15, 2012.

Dongarra, Jack (2007). Frequently Asked Questions on the Linpack Benchmark and Top500. http://www.netlib.org/utk/people/JackDongarra/faq-linpack.html (accessed August 11, 2014).

Dongarra, J.J. (2011). Performance of Various Computers Using Standard Linear Equations Software. Report CS-89-05, Computer Science Department, University of Tennessee, version of June 20, 2011.

Dupuy, Trevor N. (1976). *A Genius for War: The German Army and General Staff, 1807–1945.* Englewood Cliffs, NJ: Prentice Hall. Reprinted 1995 by Hero Books.

Echevarria, Antulio J. (1986). Auftragstaktik in Its Proper Perspective. *Military Review,* Vol. 66, No. 10, 50–56.

Echevarria, Antulio J. (1996). Moltke and the German Military Tradition: His Theories and Legacies. *Parameters,* Spring, 91–99.

Echevarria, Antulio J. (2000). *After Clausewitz: German Military Thinkers Before the Great War.* Lawrence, KS: University Press of Kansas.

English, Allan D. (1998). *Changing Face of War: Learning from History.* Montreal, Quebec: McGill-Queen's University Press.

FA53.com (2012). Mobile Phone Applications. Website for U.S. Army Mobile Phone Applications for iPhone and Android. http://www.fa53.com/apps/ (accessed February 22, 2012).

Fan, J., A. Kalyanpur, D.C. Gondek, and D.A. Ferrucci (2012). Automatic Knowledge Extraction from Documents. *IBM Journal of Research and Development,* Vol. 56, Nos. 3/4, 5:1–5:10.

Fellows, Susan, Paul Pearce, and James Moffat (2010). Measuring the Impact of Situation Awareness on Digitised Force Effectiveness. *Proc. 15th International Command and Control Research and Technology Symposium.* Santa Monica, CA, June 22–24. Washington, DC: CCRP Press. Paper 136, pp. 1–31. http://www.dodccrp.org/events/15th_iccrts_2010/papers/136.pdf (accessed August 11, 2014).

Fennell, Desmond (1988). *Investigation into the King's Cross Underground Fire.* London: Her Majesty's Stationery Office. http://www.railwaysarchive.co.uk/documents/DoT_KX1987.pdf (accessed August 11, 2014).

Fisher, C., and B. Kingma (2001). Criticality of Data Quality as Exemplified in Two Disasters. *Information and Management,* Vol. 39, No. 2, 109–116.

Fishman, Brian (2009). *Dysfunction and Decline: Lessons Learned from Inside Al Qa'ida in Iraq.* West Point, NY: Combating Terrorism Center. http://www.ctc.usma.edu/wp-content/uploads/2010/06/Dysfunction-and-Decline.pdf (accessed August 11, 2014).

Florida Governor's Disaster Planning and Response Review Committee (1992). Final Report. ("The Lewis Report" on the response to Hurricane Andrew.) Collingdale, PA: Diane Publishing. http://www.floridadisaster.org/documents/Lewis%20Report%201992.pdf.

Fox, S.G. (1995). *Unintended Consequences of Joint Digitization.* Newport, RI: U.S. Naval War College.

Frank, Richard B. (1990). *Guadalcanal: The Definitive Account of the Landmark Battle.* New York: Penguin Group.

Freedburg, S. (2012). Army Seeks New Network Tech for New Brigades Post Afghanistan, *AOL Defense,* March 19. http://defense.aol.com/2012/03/19/army-seeks-new-network-tech-for-new-brigades-post-afghanistan-m (accessed August 11, 2014).

Fuhrmann, Matthew C., Nathan D. Edwards, and Michael D. Salomone (2005). The German Offensive of 1914: A New Perspective. *Defense and Security Analysis,* Vol. 21, No. 1, 37–66.

Gardner, Nicholas (2009). Command and Control in the "Great Retreat" of 1914: The Disintegration of the British Cavalry Division. *Journal of Military History,* Vol. 63, No. 1, pp. 29–54.

Gartner (2014). Gartner Says Annual Smartphone Sales Surpassed Sales of Feature Phones for the First Time in 2013. Gartner Group, February 13. http://www.gartner.com/newsroom/id/2665715 (accessed August 11, 2014).

Gass, Gregory P. (1992). *Command and Control: The Achilles Heel of the Iran Hostage Rescue Mission.* Newport, RI: Naval War College. http://www.dtic.mil/dtic/tr/fulltext/u2/a249903.pdf (accessed August 11, 2014).

Germany (1933). *Die Truppenführung.* Berlin: Chef der Heeresleitung.

Germany (1998). *Army Command and Control.* Bonn: German Army Regulation (AR) 100/100 (Restricted).

Goodwin, Grenville (1935). The Social Divisions and Economic Life of the Western Apache. *American Anthropologist, New Series,* Vol. 37, No. 1, Part 1, 55–64.

Grant, Tim (2006). Measuring the Potential Benefits of NCW: 9/11 as Case Study. *Proc. 11th International Command and Control Research and Technology Symposium,* Cambridge, England, September 26–28. Washington, DC: CCRP Press. Paper 103, pp. 1–14. http://www.dodccrp.org/events/11th_ICCRTS/html/papers/103.pdf (accessed August 11, 2014).

Green, Dan (2009). Universal Core Improving Information Sharing Across the Government. *CHIPS,* Vol. XXVII, No. III, 23.

Greene Computing (2012). LINPACK Top 10. http://www.greenecomputing.com/apps/linpack/linpack-top-10/ (accessed August 11, 2014).

Guardian [Newspaper] (2011). 7/7 Rescue Operation Hampered by Poor Radio Communications, Inquest Hears. February 8. http://www.guardian.co.uk/uk/2011/feb/08/7july-inquest-rescue-operation-radio-communication (accessed August 11, 2014).

Hairston, W. David, Jesse Chen, Michael Barnes, Ivan Martinez, Michael Lafiandra, Mary Binseel, Angelique Scharine, Mark Ericson, Barry Vaughan, Brent Lance, and Kaleb Mcdowell (2012). Technological Areas to Improve Soldier Decisiveness: Insights from the Soldier-System Design Perspective. Technical Note ARL-TN-475. Adelphi, MD: U.S. Army Research Laboratory.

Hajkowicz, Stefan, Hannah Cook, and Anna Littleboy (2012). *Our Future World: Global Megatrends That Will Change the Way We Live. The 2012 Revision.* Canberra, Australia: Commonwealth Scientific and Industrial Research Organisation (CSIRO). http://www.csiro.au/~/media/CSIROau/Images/Other/Futures/OurFutureWorld_CSIRO_2012.pdf (accessed August 11, 2014).

Hamel, Gary (2011). First, Let's Fire all the Managers. *Harvard Business Review,* Vol. 89, No. 12, 48–60.

Harvard, James W. (2013). Airmen and Mission Command. *Air and Space Power Journal,* March–April, 131–146.

HBR (2010). How Hierarchy Can Hurt Strategy Execution. *Harvard Business Review,* Vol. 88, No. 7, 74–75.

Hidden, Andrew (1989). *Investigation into the Clapham Junction Railway Accident.* London: Her Majesty's Stationery Office. http://www.weather-charts.org/railway/Clapham_Junction_Collision_1988.pdf.

Higgins, Sarah (2008). The DCC Curation Lifecycle Model. *International Journal of Digital Curation,* Vol. 3, No. 1, 134–140. http://www.ijdc.net/index.php/ijdc/article/viewFile/69/48 (accessed August 11, 2014).

Hillsborough Independent Panel (2012). Hillsborough: The Report of the Hillsborough Independent Panel. Report HC581. London: The Stationery Office. http://hillsborough.independent.gov.uk/repository/report/HIP_report.pdf (accessed August 11, 2014).

Hodges, Harold Winter, and Edward Arthur Hughes (1936). *Select Naval Documents.* New York: Cambridge University Press. Reprinted 2009.

Hoffman, Bruce (2006). *Inside Terrorism,* 2nd ed. New York: Columbia University Press.

Hofmann, George (2006). *Through Mobility We Conquer: The Mechanization of U.S. Cavalry.* Lexington, KY: University Press of Kentucky.

Holloway, J.L. (1980). *[Iran Hostage] Rescue Mission Report.* Washington, DC: Office of the Chairman of the Joint Chiefs of Staff. http://www.history.navy.mil/library/online/hollowayrpt.htm (accessed August 11, 2014).

Hone, Trent (2006). "Give Them Hell!": The US Navy's Night Combat Doctrine and the Campaign for Guadalcanal. *War in History,* Vol. 13, No. 2, 171–199.

Howard, Courtney (2013). UAV command, control and communications. *Military and Aerospace Electronics,* Vol. 24, No. 7. http://www.military-aerospace.com/articles/print/volume-24/issue-7/special-report/uav-command-control-communications.html (accessed August 11, 2014).

Howieson, W.B. (2012). Mission Command: A Leadership Philosophy for the Health and Social Care Act 2012? *International Journal of Clinical Leadership.* Vol. 17, 217–225.

Hsu, Jeremy (2011). Military Battles Information Overload from Robot Swarms, *Tech News Daily,* September 7.

Huber, Reiner, Tor Langsaeter, Petra Eggenhofer, Fernando Freire, Antonio Grilo, Anne-Marie Grisogono, Jose Martine, Jens Roemer, Mink Spaans, and Klaus Titze (2008). The Indian Ocean Tsunami. A Case Study Investigation by NATO RTO SAS-065 Part II: The Case of Aceh and North Sumatra. Washington, DC: U.S. Department of Defense, Command and Control Research Program. http://www.dodccrp.org/files/case_studies/Tsunami_case_study.pdf (accessed August 11, 2014).

Hughes, Daniel J. (1986). Abuses of German Military History. *Military Review,* Vol. 66, No. 12, 66–76.

Hughes, Daniel J. (1993). *Moltke on the Art of War: Selected Writings.* New York: Presidio Press.

Hughes, Daniel J. (1995). Schlichting, Schlieffen, and the Prussian Theory of War in 1914. *Journal of Military History,* Vol. 59, No. 2, 257–278.

IAIDQ (2014). Information Quality. Entry in *IAIDQ Glossary*. Baltimore, MD: International Association for Information and Data Quality. http://iaidq.org/main/glossary.shtml#I (accessed August 11, 2014).

Iannotta, Ben (2012). Top Secret Goes Mobile. *Defense News*, March 29. http://www.defensenews.com/article/20120329/C4ISR02/303290008/Cover-Story-Top-Secret-Goes-Mobile (accessed August 11, 2014).

IBM, Organizational Design Community, Université Paris 1, and Aarhus University (2013). Unleashing the Potential of Big Data. White Paper based on the 2013 World Summit on Big Data and Organization Design. http://icoa.au.dk/fileadmin/ICOA/BDOD2013/Whitepaper_bigdata_final.pdf (accessed August 11, 2014).

Independent International Fact-Finding Mission on the Conflict in Georgia (2009) ["Tagliavini Report"]. *Independent International Fact-Finding Mission on the Conflict in Georgia, Report, Vol. II*. Brussels: Council of the European Union. http://www.ceiig.ch/Report.html (accessed August 11, 2014).

InnoNews (2012) (Innovation News Daily Staff). U.S. Military Bets on "Big Data" to Win Wars. *Tech News Daily*, March 29. http://www.livescience.com/19380-military-bets-big-data.html (accessed August 16, 2014).

Intel Corporation (2014). *What Happens in an Internet Minute*. Santa Clara, CA. http://www.intel.com/content/www/us/en/communications/internet-minute-infographic.html (accessed August 11, 2014).

ISO (2009). Industrial Data. ISO Technical Committee TC 184, subcommittee SC 4, ISO/TS 8000-100:2009 Data Quality—Part 100: Master Data (Overview). Geneva, Switzerland: International Standardization Organization.

ISO/IEC (1993). Information Technology—Vocabulary—Part 1: Fundamental Terms. ISO/IEC 2382-1:1993. Geneva, Switzerland: International Standardization Organization.

ISO/IEC (2008). Software Product Quality Requirements and Evaluation (SQuaRE)—Data Quality Model. Technical Committee JTC-1/SC 7, S.a.S.E., ISO/IEC 25012:2008 Software Engineering. Geneva, Switzerland: International Standardization Organization.

ITU (2003). Framework and Overall Objectives of the Future Development of IMT-2000 and Systems beyond IMT-2000. Geneva, Switzerland: International Telecommunications Union, Recommendation ITU-RM.1645.

ITU (2008). Requirements Related to Technical Performance for IMT-Advanced Radio Interface(s). Geneva, Switzerland: International Telecommunications Union, report ITU-RM.2134. http://www.itu.int/dms_pub/itu-r/opb/rep/R-REP-M.2134-2008-PDF-E.pdf (accessed February 21, 2012).

ITU (2011). *Yearbook of Statistics: Telecommunication/ICT Indicators 2001–2010*. Geneva, Switzerland: International Telecommunications Union.

ITU (2012). ICT Data and Statistics (IDS). Website of the International Telecommunication Union, http://www.itu.int/ITU-D/ict/statistics/.

ITU (2013a). *Measuring the Information Society.* Geneva, Switzerland: International Telecommunications Union.

ITU (2013b). *ITU Statistical Database.* Geneva, Switzerland: International Telecommunication Union.

Internet World Stats (2012). Facebook Statistics by Geographic Regions and World Countries. http://www.internetworldstats.com/facebook.htm (accessed February 21, 2012).

Internet World Stats (2014). Internet Growth Statistics. http://www.internet-worldstats.com/emarketing.htm.

Jacko, Julie A. (2012). *The Human–Computer Interaction Handbook,* 3rd ed. Boca Raton, FL: CRC Press.

Jackson, Brian A. (2006). Groups, Networks, or Movements: A Command-and-Control-Driven Approach to Classifying Terrorist Organizations and Its Application to Al Qaeda. *Studies in Conflict and Terrorism,* Vol. 29, 241–262.

Jaques, Elliott (1990). In Praise of Hierarchy. *Harvard Business Review,* Vol. 68, Nos. 1 and 2, 127–133.

Johnston, P. (2000). Doctrine Is not Enough: The Effect of Doctrine on the Behavior of Armies. *Parameters, U.S. Army War College Quarterly,* Autumn 2000, 30–39.

Joint Integrated Test Center (JITC) (2012). Joint Integrated Test Center Joint Tactical Radio System. United States Department of Defense, JITC.

Jones, Seth G. (2007). Fighting Networked Terrorist Groups: Lessons from Israel. *Studies in Conflict and Terrorism,* Vol. 30, No. 4, 281–302.

Jordan, Javier, Fernando M. Manas, and Nicola Horsburgh (2008). Strengths and Weaknesses of Grassroot Jihadist Networks: The Madrid Bombings. *Studies in Conflict and Terrorism,* Vol. 31, No. 1, 17–39.

Jordan, Larry R. (2008). *Hybrid War: Is the U.S. Army Ready for the Face of 21st Century Warfare?* Fort Leavenworth, KS: U.S. Army Command and General Staff College.

Kagan, Mark (1999). Redesigned Communication Equipment Strengthens First-to-Fight Operations. *AFCEA Signal Magazine Online.* http://www. afcea.org/content/?q=node/923 (accessed August 12, 2014).

Kamphuis, P.H., and H. Amersfoort (Eds.) (2010). *History of Warfare, Volume 57: May 1940: The Battle for the Netherlands.* Boston, MA: Brill Academic.

Kaut, Charles (1974). The Clan System as an Epiphenomenal Element of Western Apache Social Organization. *Ethnology,* Vol. 13, No. 1, 45–70.

Kenyon, F.G. (1910). *The Nelson Memorandum.* Tunbridge Wells, UK: J. Newns.

Kenyon, Henry (2011). New Surveillance System Offers Wide-Angle View of Battlefield. *Defense Systems,* January 3. http://defensesystems.com/articles/2011/01/03/air-force-gorgon-stare-wide-angle-surveillance.aspx.

Kenyon, Henry (2012). Marines Want Smart Phone for Classified, Commercial Systems. *GCN,* April 2. http://gcn.com/Articles/2012/04/02/Marine-Corps-launches-trusted-mobile-device-program.aspx?Page = 1 (accessed August 11, 2014).

Khalilzad, Z., J. White, and A.W. Marshall (1999). *Strategic Appraisal: The Changing Role of Information in Warfare*. Santa Monica, CA: RAND Corporation, http://www.rand.org/pubs/monograph_reports/MR1016.

King, Ernest (1941). Exercise of Command—Excess of Detail in Orders and Instructions. U.S. Commander in Chief, Atlantic Command (CINCLANT) Serial 053, January 21. http://blog.usni.org/2012/08/29/the-wisdom-of-a-king (accessed August 11, 2014).

King, Ernest (1944). *War Instructions United States Navy 1944*. Washington, DC: U.S. Fleet, Headquarters of the Commander in Chief, Navy Department. http://www.history.navy.mil/library/online/war_instruct.htm (accessed August 11, 2014).

Kline (1985). *Research, Invention, Innovation and Production: Models and Reality*. Report INN-1, March 1985, Mechanical Engineering Department, Stanford University. Stanford, CA: Stanford University.

Kline, S.J. and N. Rosenberg (1986). An Overview of Innovation. In R. Landau & N. Rosenberg (Eds.), *The Positive Sum Strategy: Harnessing Technology for Economic Growth*. Washington, DC: National Academy Press, pp. 275–305.

Knabb, Richard D., Jamie R. Rhome, and Daniel P. Brown (2005). *Tropical Cyclone Report: Hurricane Katrina, August 23-30, 2005*. Miami, FL: National Hurricane Center. http://www.nhc.noaa.gov/pdf/TCR-AL122005_Katrina.pdf (accessed August 11, 2014).

Knox, Dudley W. (1913). Trained Initiative and Unity of Action: The True Bases of Military Efficiency. *U.S. Naval Institute Proceedings*, Vol. 39, No. 1, 41–62.

Knox, Dudley W. (1914a). The Great Lesson from Nelson for Today. *U.S. Naval Institute Proceedings*, Vol. 40, No. 2, 295–318.

Knox, Dudley W. (1914b). Old Principles and Modern Applications. *U.S. Naval Institute Proceedings*, Vol. 40, No. 4, 1009–1039.

Knox, Dudley W. (1915). The Role of Doctrine in Naval Warfare. *U.S. Naval Institute Proceedings*, Vol. 41, No. 2, 325–354.

Komorowski, Matt (2011). A History of Storage Cost. http://www.mkomo.com/cost-per-gigabyte (accessed August 11, 2014).

Kotter, John P. (2012). Accelerate! *Harvard Business Review*, Vol. 90, No. 11, 44–58.

Kramer, Corinne, Marius Vassiliou, and Jonathan R. Agre (2012). Keyword Analysis of Command and Control-Related Science and Technology Efforts of the United States Department of Defense. *Proc. 17th International Command and Control Research and Technology Symposium*, Fairfax, VA, June 19–21. Washington, DC: CCRP Press. Paper 067, pp. 1–18. http://www.dodccrp.org/events/17th_iccrts_2012/post_conference/papers/067.pdf (accessed August 11, 2014).

Krulak, Charles C. (1999). The Strategic Corporal: Leadership in the Three Block War. *Marines Magazine*, Vol. 28, No. 1, 32.

Kugler, Richard (2007). *Operation Anaconda in Afghanistan: A Case Study of Adaptation in Battle (Case Studies in Defense Transformation, Number 5)*. Washington, DC: National Defense University, Center for Technology and National Security Policy.

Kugler, R., M. Baranick, and H. Binnendijk (2009). *Operation Anaconda: Lessons for Joint Operations*. Washington, DC: National Defense University, Center for Technology and National Security Policy.

Kwoh, Leslie (2011). Smartphones to Overtake Traditional Cell Phones, become the new "Standard." *New Jersey Star-Ledger*, 4 September. http://www.nj.com/business/index.ssf/2011/09/smartphones_overtake_feature_p.html (accessed August 11, 2014).

Lazer, David, Ryan Kennedy, Gary King, and Alessandro Vespignani (2014). The Parable of Google Flu: Traps in Big Data Analysis. *Science*, Vol. 343, No. 6176, 1203–1205.

Leistenschneider, Stephan (2002). Auftragstaktik im preußisch-deutschen Heer 1871 bis 1914. Hamburg: E.S. Mittler and Sohn.

Leslie, Andrew, Peter Gizewski, and Michael Rostek (2008). Developing a Comprehensive Approach to Canadian Forces Operations. *Canadian Military Journal*, Vol. 9, No. 1, 11–20.

Lieberman, Charles A., and Serguei Cheloukhine (2009). 2005 London Bombings. Chapter 13 in M.R. Haberfeld and Agostino Von Hassell (Eds.), *A New Understanding of Terrorism: Case Studies, Trajectories and Lessons Learned* (233–248). New York: Springer.

Lizotte, M., F. Bernier, M. Mokhtari, E. Boivin, M.B. DuCharme, and D. Poussart (2008). IMAGE: Simulation for Understanding Complex Situations and Increasing Future Force Agility. *Proc. 26th Army Science Conference*, Orlando, FL.

Lizotte, M., F. Bernier, M. Mokhtari, and E. Boivin (2013). IMAGE Final Report: An Interactive Computer-Aided Cognition Capability for C4ISR Complexity Discovery (Report No. TR 2013-397). Québec, Canada: Defence R&D Canada-Valcartier.

MacPherson, Malcolm (2006). *Roberts Ridge: A Story of Courage and Sacrifice on Takur Ghar Mountain, Afghanistan*. New York: Dell.

Makinen, Gail (2002). The Economic Effects of 9/11: A Retrospective Assessment. Report RL31617. Washington, DC: Library of Congress, Congressional Research Service. http://www.fas.org/irp/crs/RL31617.pdf (accessed August 11, 2014).

Manyika, James, Michael Chui, Brad Brown, Jacques Bughin, Richard Dobbs, Charles Roxburgh, and Angela Hung Byers (2011). Big Data: The Next Frontier for Innovation, Competition, and Productivity. New York: McKinsey Global Institute.

Martini, Antonella, Mariano Corso, and Luisa Pellegrini (2009). An Empirical Roadmap for Intranet Evolution. *International Journal of Information Management*, Vol. 29, No. 4, 295–308.

McDaniel, John R.M. (2001). C2 Case Study: The FSCL in Desert Storm. *Proc. Sensemaking Workshop*, Vienna, Virginia. Washington, DC: Command and Control Research Program. http://www.dodccrp.org/events/2001_sense-making_workshop/pdf/C2_FSCL_doc.pdf (accessed August 11, 2014).

McDermott, Roger N. (2009). Russia's Conventional Armed Forces and the Georgian War. *Parameters*, Spring, 65–80.

McGrath, Kevin (2011). *Confronting Al Qaeda, New Strategies to Combat Terrorism*. Annapolis, MD: Naval Institute Press.

McHale, J. (2012). Military Unmanned Systems, C4ISR, Avionics, and Vectronics Markets Are COTS Markets. Military Embedded Systems, September 4. http://mil-embedded.com/articles/military-avionics-vetronics-markets-cots-markets/ (accessed August 11, 2014).

Michael, George (2012). Leaderless Resistance: The New Face of Terrorism. *Defence Studies*, Vol. 12, No. 2, 257–282.

Miller, Michael (2010). Cisco: Internet Moves 21 Exabytes per Month. *PC Magazine* Online, March 25. http://www.pcmag.com/article2/0,2817,2361820,00.asp (accessed August 11, 2014).

Mintzberg, Henry (1979). *The Structuring of Organizations: A Synthesis of the Research*. Englewood Cliffs, NJ: Prentice Hall.

Mintzberg, Henry (1980). Structure in 5's: A Synthesis of the Research in Organizational Design. *Management Science*, Vol. 26, No. 3, 322–341.

Moody, Theodore J. (2010). *Filling the Gap between NIMS/ICS and the Law Enforcement Initial Response in the Age of the Urban Jihad*. Master's Thesis. Monterey, CA: Naval Postgraduate School. http://www.dtic.mil/dtic/tr/fulltext/u2/a531492.pdf (accessed August 12, 2014).

Moran, Cindy (2011). DISA Network Services. Federal Networks Conference, Washington, DC, February 15–16.

Moreau, Luc, Paul Groth, Simon Miles, Javier Vazquez-Salceda, John Ibbotson, Sheng Jiang, Steve Munroe, Omer Rana, Andreas Schreiber, Victor Tan, and Laszlo Varga (2008). The Provenance of Electronic Data. *Communications of the ACM*, Vol. 51, No. 4, 52–58.

Moynihan, Donald P. (2006). What Makes Hierarchical Networks Succeed? Evidence from Hurricane Katrina. Annual Meeting of the Association of Public Policy and Management, Madison, WI, November 2–4. http://www.lafollette.wisc.edu/appam/moynihankatrina.pdf (accessed August 11, 2014).

Moynihan, Donald P. (2009). *The Response to Hurricane Katrina*. Geneva, Switzerland: International Risk Governance Council. http://www.irgc.org/IMG/pdf/Hurricane_Katrina_full_case_study_web.pdf.

Munoz, C. (2011). Army Troops Slam New Combat Smartphone. *AOL Defense*, November 23. http://defense.aol.com/2011/11/23/armt-troops-slam-new-combat-smartphone (accessed August 11, 2014).

Murray, Williamson, and Allan R. Millett (2000). *A War to Be Won: Fighting the Second World War, 1937–1945*. Cambridge, MA: Harvard University Press.

Muth, Jorg (2011). *Command Culture: Officer Education in the U.S. Army and the German Armed Forces, 1901–1940, and the Consequences for World War II*. Denton, TX: University of North Texas Press.

Myers, S.L. (2000). Chinese Embassy Bombing: A Wide Net of Blame. *New York Times*, April 17.

Nachira, A., and G. Mazzini (2011). The Value of the Radio Spectrum: A Comparison between Theory and Auctions. *Proc. IEEE International Conference on Ultra-Wideband (ICUWB)* (pp. 312–316).

Naisbitt, John (1982). *Megatrends: Ten New Directions Transforming Our Lives.* New York: Warner Books.

National Commission on Terrorist Attacks upon the United States (2004). *The 9/11 Commission Report: Final Report of the National Commission on Terrorist Attacks upon the United States.* Washington, DC: National Commission on Terrorist Attacks upon the United States. http://www.9-11commission.gov/report/911Report.pdf (accessed August 11, 2014).

NATO (2006). Exploring New Command and Control Concepts and Capabilities. Technical Report RTO-TR-SAS-050. Brussels, Belgium: North Atlantic Treaty Organization. http://www.dodccrp.org/files/SAS-050%20Final%20Report.pdf (accessed August 11, 2014).

NATO (2008). NATO Glossary of Terms and Definitions (English and French). Publication AAP-6(2008). Brussels, Belgium: North Atlantic Treaty Organization Standardization Agency.

NATO (2012). Overview of the Joint C3 Information Exchange Data Model (JC3IEDM Overview). Greding, Germany: NATO Multilateral Interoperability Program. https://mipsite.lsec.dnd.ca/Public%20Document%20Library/04-Baseline_3.1/Interface-Specification/JC3IEDM/JC3IEDM-Overview-3.1.4.pdf (accessed August 11, 2014).

NATO (2013). C2 Agility. Technical Report STO-TR-SAS-085. Brussels, Belgium: North Atlantic Treaty Organization. http://dodccrp.org/sas-085/sas-085_report_final.pdf (accessed August 11, 2014).

Nelson, Horatio (1805). Memorandum. HMS Victory, Off Cadiz, 9th October 1805. Reproduced by the British Library at http://www.bl.uk/learning/timeline/item106127.html (accessed August 11, 2014).

Nokia Siemens Networks (2011). *Designing, Operating, and Optimizing Unified Heterogeneous Networks.* Espoo, Finland: Nokia Siemens Networks.

OECD (2011). *OECD Factbook 2011–2012: Economic, Environmental and Social Statistics.* Paris: Organization for Economic Cooperation and Development.

OECD (2012a). *OECD Communications Outlook 2011.* Paris: Organization for Economic Cooperation and Development.

OECD (2012b). *OECD Internet Economy Outlook 2012.* Paris: Organization for Economic Cooperation and Development.

Oklahoma Department of Civil Emergency Management (2003). *After Action Report: Alfred P. Murrah Federal Building Bombing, 19 April 1995 in Oklahoma City, Oklahoma.* Oklahoma City, OK: Department of Central Services Central Printing Division, Publications Clearinghouse of the Oklahoma Department of Libraries. http://www.ok.gov/OEM/documents/Bombing%20After%20Action%20Report.pdf (accessed August 11, 2014).

Oliveiro, Chuck (1998). Trust, Manoeuvre Warfare, Mission Command, and Canada's Army. *Canadian Army Journal*, Vol. 1, No. 1, 1–3.

Ophir, Eyal, Clifford Nass, and Anthony D. Wagner (2009). Cognitive Control in Media Multitaskers. *Proc. National Academy of Sciences*, Vol. 106, No. 37, 15583–15587.

Orenstein, Jack, and Marius Vassiliou (2014). Issues in "Big-Data" Database Systems. *Proc. 19th International Command and Control Research and Technology Symposium*, Alexandria, VA, June 19-21. Washington, DC: CCRP Press. Paper 113, pp. 1–18. http://www.dodccrp.org/events/19th_iccrts_2014/post_conference/papers/113.pdf (accessed August 11, 2014).

OSPA (2013). OPSEC Glossary of Terms. Operations Security Professional's Association. http://www.opsecprofessionals.org/terms.html (accessed August 11, 2014).

Ouchi, William G., and Raymond L. Price (1978). Hierarchies, Clans, and Theory Z: A New Perspective on Organization Development. *Organizational Dynamics*, Vol. 7, No. 2, 25–44.

Palmer, Michael A. (2005). *Command at Sea: Naval Command and Control since the Sixteenth Century*. Cambridge, MA: Harvard University Press.

Parasuraman, Raja (2011). Neuroergonomics: Brain, Cognition, and Performance at Work. *Current Directions in Psychological Science*, Vol. 20, No. 3, 181–186.

Parasuraman, Raja, and Glenn F. Wilson (2008). Putting the Brain to Work: Neuroergonomics Past, Present, and Future. *Human Factors*, Vol. 50, No. 3, 468–474.

Parliament of Victoria, 2009 Victorian Bushfires Royal Commission (2010). *2009 Bushfires Royal Commission Final Report*. Melbourne, Victoria, Australia: Government Printer to the State of Victoria. http://www.royalcommission.vic.gov.au/finaldocuments/summary/PF/VBRC_Summary_PF.pdf (accessed August 11, 2014).

Pearce, P., A. Robinson, and S. Wright (2003). The Wargame Infrastructure and Simulation Environment (Wise). *Proc. Knowledge-Based Intelligent Information and Engineering Systems Conference* (pp. 714–722). Oxford, UK.

Pigeau, Ross, and Carol McCann (2002). Reconceptualizing Command and Control. *Canadian Military Journal*, Vol. 3, No. 1, 53–64.

Pigeon, Stéphane, Carl Swail, Edouard Geoffrois, Cristine Bruckner, David van Leeuwen, Carlos Teixeira, Ozgur Orman, Paul Collins, Timothy Anderson, Jon Grieco, and Marc Zissman (2005). Use of Speech and Language Technology in Military Environments. RTO Technical Report TR-IST-037. Brussels, Belgium: North Atlantic Treaty Organization. http://www.stephanepigeon.com/Docs/TR-IST-037-ALL.pdf (accessed August 11, 2014).

Pipino, L.L., Y.W. Lee, and R.Y. Wang (2002). Data Quality Assessment. *Communications of the ACM*, Vol. 45, No. 4, 211–218.

Plumb, M. (2012). Fantastic 4G. *IEEE Spectrum*, Vol. 49, No. 1, 51–53.

Popham, Home (1803). *Telegraphic Signals; or Marine Vocabulary*. "Near Whitehall," England: T. Egerton, Military Library.

Prensky, M. (2001a). Digital Natives, Digital Immigrants Part 1. *On the Horizon*, Vol. 9, No. 5, 1–6.

Prensky, M. (2001b). Digital Natives, Digital Immigrants Part 2: Do They Really Think Differently? *On the Horizon*, Vol. 9, No. 6, 1–6.

Prussia (Germany) Kriegsministerium (1888a). *Exerzier-Reglement für die Infanterie*. Berlin: Ernst Siegfried Mittler und Sohn.

Prussia (Germany) War Ministry (1888b). *German Field Exercise 1888. Part II. The Fight*. Translated by Captain W. H. Sawyer. London: Edward Stanford.

Pugh, D.S., D.J. Hickson, and C.R. Hinings (1969). An Empirical Taxonomy of Structures of Work Organizations. *Administrative Science Quarterly*, Vol. 14, No. 1, 115–126.

PWC (2013). *Building Trust in a Time of Change: Global Annual Review 2013*. New York: PricewaterhouseCoopers International. http://www.pwc.com/gx/en/annual-review/2013/assets/pwc-global-annual-review-2013.pdf (accessed August 11, 2014).

R&D (2012). 2013 Global R&D Funding Forecast. *R&D Magazine*, December, 1–35.

Ram, S., and J. Liu (2008). A Semiotics Framework for Analyzing Data Provenance Research. *Journal of Computing Science and Engineering*, Vol. 2, No. 3, 221–248.

Ramshaw, Ryan (2007). C2-Less Is More. *Proc. 12th International Command and Control Research and Technology Symposium* (June 19–21, Newport, RI). Washington, DC: CCRP Press, Paper 003 pp. 1–8. http://www.dodccrp.org/events/12th_ICCRTS/Papers/003.pdf (accessed August 11, 2014).

Rappaport, Ed (2005). Preliminary Report: Hurricane Andrew. (Updated). Miami, FL: National Hurricane Center. http://www.nhc.noaa.gov/1992andrew.html (accessed August 11, 2014).

Reily, Todd, and Martina Balestr (2011). Applying Gestural Interfaces to Command-and-Control, Design, User Experience and Usability: Theory, Methods, Tools and Practice, First International Conference, DUXU 2011, Held as Part of HCI International 2011, Orlando, FL, July 9–14, *Proceedings*, Part II, Ed. Aaron Marcus (pp. 187–194). Berlin-Heidelberg: Springer.

Richtel, Matt (2010a). Attached to Technology and Paying a Price. *New York Times*, June 6.

Richtel, Matt (2010b). Digital Devices Deprive Brain of Needed Downtime. *New York Times*, August 24.

Rittel, Horst W.J., and Melvin M. Webber (1973). Dilemmas in a General Theory of Planning. *Policy Sciences*, Vol. 4, No. 2, 155–169.

Robinson, Brian (2009). New UAV Sensors Could Leave Enemy No Place to Hide. *Defense Systems*, September 9. http://defensesystems.com/articles/2009/09/02/c4isr-3-gorgon-stare.aspx (accessed August 11, 2014).

Rogers, Yvonne, Helen Sharp, and Jenny Preece (2011). *Interaction Design: Beyond Human-Computer Interaction*. 3rd ed. New York: John Wiley and Sons.

Rourke, Kellie S. (2009). *U.S. Counterinsurgency Doctrine: Is It Adequate to Defeat Hezbollah as a Threat Model of Future Insurgencies?* Fort Leavenworth, KS: U.S. Army Command and General Staff College.

Ruddy, Mary (2007). ELICIT—The Experimental Laboratory for the Investigation of Collaboration, Information Sharing, and Trust. *Proc. 12th International Command and Control Research and Technology Symposium*,

Newport, RI, June 19-21. Washington, DC: CCRP Press, Paper 155 pp. 1–74. http://dodccrp.org/events/12th_ICCRTS/CD/html/papers/155. pdf (accessed August 11, 2014).

Ruppel, Cynthia P., and Susan J. Harrington (2001). Sharing Knowledge through Intranets: A Study of Organizational Culture and Intranet Implementation. *IEEE Transactions on Professional Communication*, Vol. 44, No. 1, 37–52.

Saunders, Clayton D. (2002). *Al Qaeda: An Example of Network-Centric Operations*. Newport, RI: U.S. Naval War College.

Schwartz, Carey (2011). Data-to-Decisions S&T Priority Roadmap Briefing. Arlington, VA: Office of Naval Research. http://www.acq.osd.mil/chieftechnologist/publications/docs/2011%2011%207%20Data%20to%20 Decisions%20PSD%20Roadmap.pdf (accessed August 11, 2014).

SDSS (2014). The Sloan Digital Sky Survey. http://www.sdss.org/ (accessed August 11, 2014).

Sebastian, Mike (2008). The Cost of Dirty Data to Accounts Receivable Managers. Inside Accounts Receivable Management, insideARM.com. http://www. insidearm.com/daily/collection-technologies/collection-technology/the-cost-of-dirty-data-to-accounts-receivable-managers/ (accessed August 11, 2014).

Seddon, John (2005). *Freedom from Command and Control*. New York: Productivity Press.

Senn, Matthew A., and James D. Turner (2008). *Analysis of Satellite Communication as a Method to Meet Information Exchange Requirements for the Enhanced Company Concept*. Monterey, CA: Naval Postgraduate School.

Shamir, Eitan (2011). *Transforming Command: The Pursuit of Mission Command in the U.S., British, and Israeli Armies*. Stanford, CA: Stanford University Press.

Shankaranarayan, G., M. Ziad, and R. Wang (2003). Managing Data Quality in Dynamic Decision Environments: An Information Product Approach. *Journal of Database Management*, Vol. 14, No. 4, 14–32.

Shanker, Thom, and Matt Richtel (2011). In New Military, Data Overload Can Be Deadly. *New York Times*, January 16. http://www.nytimes. com/2011/01/17/technology/17brain.html?pagewanted = all&_r = 0 (accessed August 11, 2014).

Shannon, Claude (1949). Communication in the Presence of Noise. *Proc. Institute of Radio Engineers*, Vol. 37, 10–21. Reprinted in D. Slepian, Ed., *Key Papers in the Development of Information Theory*, IEEE Press, New York, 1974. Reprinted in *Proc. Institute of Electrical and Electronic Engineers*, Vol. 72 (1984), pp. 1192–1201. Included in Part A.

Shariat, Sheryll, Sue Mallonee, and Shelli Stephens-Stidham (1998). *Summary of Reportable Injuries in Oklahoma: Oklahoma City Bombing Injuries*. Oklahoma City, OK: Oklahoma State Department of Health. http:// web.archive.org/web/20080110063748/http://www.health.state.ok.us/ PROGRAM/injury/Summary/bomb/OKCbomb.htm (accessed August 11, 2014).

Shields, Christopher A. (2012). *American Terrorist Trials, Prosecutorial and Defense Strategies*. El Paso, TX: LFB Scholarly.

Shumberger, Michael, Andrew Duchowski, and Kevin Charlow (2005). Human Factors Engineering in Command, Control and Intelligence (C2I) User Interface Development Processes: Use of Eye Tracking Methodology. *Proc. Human Systems Integration Symposium*, American Society of Naval Engineers (ASNE), June 20–22, Arlington, VA. Alexandria, VA: American Society of Naval Engineers, pp. 1–13. http://andrewd.ces.clemson.edu/research/vislab/docs/ASNE-HSI-05.pdf (accessed August 11, 2014).

Simon, A.J. (2006). Overview of the Department of Defense Net-Centric Strategy. *Crosstalk, Journal of Defense Software Engineering*. Vol. 19, No. 7, 21–22.

Singh, Sarwant (2011). *Throwing Light on the Future: Megatrends That Will Shape the World*. London: Frost and Sullivan, Visionary Innovation Research Group. http://www.gilcommunity.com/files/6413/6252/4228/Sarwant_Singh_-_Frost__Sullivan.pdf (accessed August 11, 2014).

Smith, Donald Steven (2006). U.S. Must Network to Defeat al Qaeda, Kimmitt Says. Washington, DC: American Forces Press Service, February 21.

Soknich, William (2009). Strategic Knowledge Integration Web (SKIWeb)—Global Awareness Presentation Services (GAPS). *Proc. 2009 Commercial Joint Mapping ToolKit (CJMTK) Annual Conference*, April 7. McLean, VA: Northrop Grumman Information Systems, pp. 1–21. http://www.cjmtk.com/EventRegistration/CjmtkConf09/CJMTK_UC_2009_SKIWEB-GAPS.PDF (accessed August 11, 2014).

Stewart, Keith (2006). Coalition Command and Control in the Networked Era. *Proc. 11th International Command and Control Research and Technology Symposium* (June 20-22, Cambridge, UK). Washington, DC: CCRP Press. Paper 026, pp. 1-23. http://dodccrp.org/events/11th_ICCRTS/html/papers/026.pdf (accessed August 11, 2014).

Stewart, Keith (2009). Command Approach: Problem Bounding vs. Problem Solving in Mission Command. *Proc. 14th International Command and Control Research and Technology Symposium* (June 15–17, Washington, DC). Washington, DC: Washington, DC: CCRP Press. Paper 176, pp. 1–8. http://dodccrp.org/events/14th_iccrts_2009/papers/176.pdf (August 11, 2014).

Stoker, L. (2012). Battlefield Smartphones Receive a Ringing Endorsement. ArmyTechnology.com, July 31. http://www.army-technology.com/features/featurebattlefield-smartphones-endorsement-technology (accessed August 11, 2014).

Strong, D.M., Y.W. Lee, and R.Y. Wang (1997). Data Quality in Context. *Communications of the ACM*, Vol. 40, No. 5, 103–110.

Stulberg, Adam (2009). Organizing for Revolutionary Effect: Managing the Many Faces of Network-Centric Operations. *Proc. International Studies Association Annual Convention*, New York, February 15–18.

Takai, Teresa M. (2013). Adoption of the National Information Exchange Model within the Department of Defense. Memorandum, March 28. Washington, DC: U.S. Department of Defense, Office of the Chief Information Officer. http://dodcio.defense.gov/Portals/0/

Documents/2013-03-28%20Adoption%20of%20the%20NIEM%20 within%20the%20DoD.pdf (accessed August 11, 2014).

Tavis, Anna, Richard Vosburgh, and Ed Gubman (Eds.) (2012). *Point Counterpoint: New Perspectives on People and Strategy.* Alexandria, VA: Society for Human Resource Management.

Taylor, Frederick Winslow (1911). *The Principles of Scientific Management.* New York: Harper and Brothers (1913 reprint).

Thomas, Charles S. (1987). The Iranian Hostage Rescue Attempt. Report AD-A183-395. Carlisle, PA: U.S. Army War College. http://www.dtic.mil/cgi-bin/GetTRDoc?AD=ADA183395 (accessed August 11, 2014).

Toal, Mark J. (1998). *The Mayaguez Incident: Near Disaster at Koh Tang.* Quantico, VA: Marine Corps War College. http://www.dtic.mil/cgi-bin/GetTRDoc?AD=ADA529638 (accessed August 11, 2014).

Tucker, Spencer C. (Ed.) (2010). *Encyclopedia of Middle East Wars.* Denver, CO: ABC Clio.

Tyler, Barbara Ann (1965). Cochise: Apache War Leader, 1858–1861. *Journal of Arizona History,* Vol. 6, No. 1, 1–10.

U.K. Army (2005). *Land Operations.* Shrivenham, UK: United Kingdom Ministry of Defence, Director General, Development, Concepts, and Doctrine, Publication AC 71819.

U.K. Ministry of Defence (1989). Design for Military Operations. Shrivenham, UK: United Kingdom Ministry of Defence, Director General, Development, Concepts, and Doctrine, Publication AC 71451.

U.K. Royal Air Force (2008). British Air and Space Power Doctrine (AP3000), 4th ed. Cranwell, Lincolnshire, UK: Royal Air Force Center for Air Power Studies.

Uma, M., and G. Padmavathi (2013). A Survey of Various Cyber Attacks and Their Classification. *International Journal of Network Security,* Vol. 15, No. 6, 391–397.

United Nations (2012). Composition of Macro Geographical (Continental) Regions, Geographic Sub-Regions, and Selected Economic and Other Groupings (UN M49). http://unstats.un.org/unsd/methods/m49/m49regin.htm (accessed August 11, 2014).

U.S. Air Force (2007). Command and Control. Washington, DC: U.S. Air Force, Doctrine Document 2-8.

U.S. Army (2003). Mission Command: Command and Control of Army Forces. Washington, DC: Headquarters, U.S. Department of the Army, Field Manual No. 6-0.

U.S. Army (2005). Command and Control Information Exchange Data Model. DAMO-SSB Memo 28. Washington, DC: U.S. Department of the Army, Military Operations.

U.S. Army (2010). Army Net-Centric Data Strategy (ANCDS). Washington, DC: U.S. Army. https://secureweb2.hqda.pentagon.mil/vdas_army-posturestatement/2010/information_papers/Army_Net-Centric_Data_Strategy_%28ANCDS%29.asp (accessed August 11, 2014).

U.S. Conference of Mayors (2004). *The United States Conference of Mayors Interoperability Survey: A 192-City Survey*. Washington, DC: U.S. Conference of Mayors. http://usmayors.org/72ndAnnualMeeting/interoperabilityreport_062804.pdf (accessed August 11, 2014).

U.S. Department of Defense (2014). Department of Defense Dictionary of Military and Associated Terms. Joint Publication 1-02. Washington, DC: U.S. Department of Defense.

U.S. Fire Administration (1999). Wanton Violence at Columbine High School, Littleton, Colorado. Report USFA-TR-128. Emmitsburg, MD: U.S. Fire Administration. http://www.usfa.fema.gov/downloads/pdf/publications/tr-128.pdf (accessed August 11, 2014).

U.S. Government (2014). National Information Exchange Model. U.S. Government. https://www.niem.gov/Pages/default.aspx.

U.S. Government Accountability Office (1976). The Seizure of the Mayaguez: A Case Study of Crisis Management. Report B-133001. Washington, DC: Government Accountability Office [then called General Accounting Office]. http://www.gao.gov/assets/240/233910.pdf (accessed August 11, 2014).

USGS (2013). Earthquakes with 1,000 or More Deaths since 1900. http://earthquake.usgs.gov/earthquakes/world/world_deaths.php (accessed August 11, 2014).

U.S. House of Representatives (2006). A Failure of Initiative: Final Report of the Select Bipartisan Committee to Investigate the Preparation for and Response to Hurricane Katrina. U.S. House of Representatives, 109th Congress, 2nd Session, Report 000-000. Washington, DC: U.S. Government Printing Office. http://katrina.house.gov/full_katrina_report.htm (accessed August 11, 2014).

USJCS (2012). Information Operations. Joint Publication 3-13. Washington, DC: U.S. Department of Defense, Joint Chiefs of Staff. http://www.dtic.mil/doctrine/new_pubs/jp3_13.pdf (accessed August 11, 2014).

USJS J-7 (2008). Joint Capability Areas: JCA 101. Washington, DC: U.S. Department of Defense, Joint Staff. http://www.dtic.mil/futurejointwarfare/strategic/jca101.ppt (accessed August 11, 2014).

USJS J-7 (2014). Joint Capability Areas. Washington, DC: U.S. Department of Defense. http://www.dtic.mil/futurejointwarfare/jca.htm (accessed August 11, 2014).

USMC (1989). Warfighting. Washington, DC: Department of the Navy, Headquarters, U.S. Marine Corps, Publication FMFM 1.

USMC (1996). Command and Control. Washington, DC: Department of the Navy, Headquarters, U.S. Marine Corps, Doctrine Publication MCDP 6.

USMC (2011). Pre-Solicitation for Trusted Handheld Platform. No. M6785412I2414, November 11. Washington, DC: U.S. Marine Corps.

USONI (1943). *Combat Narratives: Solomon Islands Campaign II: The Battle of Savo Island, 9 August 1942*. Washington, DC: U.S. Navy, Office of Naval Intelligence. http://www.history.navy.mil/library/online/battlesavoisland1942.htm (accessed August 11, 2014).

U.S. Senate (2006). Hurricane Katrina: A Nation Still Unprepared. Special Report of the Committee on Homeland Security and Governmental Affairs, U.S. Senate, Together with Additional Views. 109th Congress, 2nd Session, Report S. Rept. 109-322. Washington, DC: U.S. Government Printing Office. http://www.gpo.gov/fdsys/pkg/CRPT-109srpt322/pdf/CRPT-109srpt322.pdf (accessed August 11, 2014).

Van Creveld, Martin L. (1985). Command in War. Cambridge, MA: Harvard University Press.

van de Ven, Andrew H. (1976). A Framework for Organization Assessment.

Vassiliou, Marius, V. Sundareswaran, S. Chen, R. Behringer, C. Tam, M. Chan, P. Bangayan, and J. McGee (2000). Integrated Multimodal Human-Computer Interface and Augmented Reality for Interactive Display Applications. In Darrel G. Hopper (Ed.), *Cockpit Displays VII: Displays for Defense Applications (Proc. SPIE 4022)*, 106–115.

Vassiliou, Marius (2010). The Evolution Towards Decentralized C2. *Proc. 15th International Command and Control Research and Technology Symposium (ICCRTS)*, Santa Monica, CA, June 22-24. Washington, DC: CCRP Press. Paper 054, pp. 1–23. http://www.dodccrp.org/events/15th_iccrts_2010/papers/054.pdf (accessed August 11, 2014).

Vassiliou, Marius, S.O. Davis, and Jonathan Agre (2011). Innovation Patterns in Some Successful C2 Technologies. *Proc. 16th International Command and Control Research and Technology Symposium (ICCRTS).*, Québec City, Québec, Canada, June 21–23. Washington, DC: CCRP Press. Paper 030, pp. 1–20. http://www.dodccrp.org/events/16th_iccrts_2011/papers/030.pdf (accessed August 11, 2014).

Vassiliou, Marius, and David S. Alberts (2012). Megatrends Reshaping C2 and their Implications for Science and Technology Priorities. *Proc. 17th International Command and Control Research and Technology Symposium*, Fairfax, VA, June 19–21. Washington, DC: CCRP Press. Paper 066, pp. 1–23. http://www.dodccrp.org/events/17th_iccrts_2012/post_conference/papers/066.pdf (accessed August 12, 2014).

Vassiliou, Marius, and David S. Alberts (2013). C2 Failures: A Taxonomy and Analysis. *Proc. 18th International Command and Control Research and Technology Symposium (ICCRTS)*, Alexandria, VA, June 19–21. Washington, DC: CCRP Press. Paper 049, pp. 1–25. http://www.dodccrp.org/events/18th_iccrts_2013/post_conference/papers/049.pdf (accessed August 17, 2014).

Vassiliou, Marius, Jonathan R. Agre, Syed Shah, and Thomas MacDonald (2013). Crucial Differences between Commercial and Military Communications Technology Needs: Why the Military Still Needs Its Own Research. *Proc. IEEE Military Communications Conference MILCOM 13* (pp. 342–347).

Vincent, Edgar (2003). Nelson and Mission Command. *History Today*, Vol. 53, No. 6, 18–19.

Vision Mobile (2011). *Mobile Platforms: The Clash of Ecosystems*. London: Vision Mobile.

Vogelaar, L.W., and Eric Hanskramer (2004). Mission Command in Dutch Peace Support Missions. *Armed Forces & Society*, Vol. 30, No. 3, 409–431.

Walker, Guy H., Neville A. Stanton, Paul M. Salmon, and Daniel P. Jenkins (2009). *Command and Control: The Sociotechnical Perspective*. Burlington, VT: Ashgate.

Walsh, Philip, Stanley O. Davis, Maryann Kiefer, Kyle A. Morrison, James W. Pipher, and Henry G. Potrykus (2009). DoD Net-Centric Services Strategy Implementation in the C2 Domain. IDA Paper P-4549. Alexandria, VA: Institute for Defense Analyses.

Waters, Jeff, Brenda J. Powers, and Marion G. Ceruti (2009). Global Interoperability Using Semantics, Standards, Science and Technology (GIS3T). *Computer Standards and Interfaces,* Vol. 31, No. 6, 1158–1166.

Watt, Robert N. (2011). Victorio's Military and Political Leadership of the Warm Springs Apaches. *War in History*, Vol. 18, No. 4, 457–494.

Wawro, Geoffrey (2005). *The Franco-Prussian War: The German Conquest of France in 1870–1871*. New York: Cambridge University Press.

White, Tom (2012). *Hadoop: The Definitive Guide*. Sebastopol, CA: O'Reilly Media.

Winter, Jay (2010). Demography. Chapter 17 in John Horne, Ed., *Companion to World War I*. New York: Blackwell (pp. 248–262).

Wittmann, Jochen (2012). *Auftragstaktik*. Berlin: Carola Hartmann Miles Verlag.

Wolfram, Stephen (2010). Making the World's Data Computable. Keynote Speech at the Wolfram Data Summit, Washington, DC, September 9. http://blog. stephenwolfram.com/2010/09/making-the-worlds-data-computable/ (accessed August 11, 2014).

Wolfram|Alpha (2014). Making the World's Knowledge Computable. Champaign, IL: Wolfram Research. http://www.wolframalpha.com/about.html (accessed August 11, 2014).

Wyld, David C. (2007). *The Blogging Revolution: Government in the Age of Web 2.0*. Washington, DC: IBM Center for the Business of Government.

Wyly, M.D. (1991). Thinking Like Marines. https://fasttransients.files.word-press.com/2012/03/wyly-thinking-like-marines.pdf (accessed August 11, 2014).

Yardley, Ivan, and Andrew Kakabadse (2007). Understanding Mission Command: A Model for Developing Competitive Advantage in a Business Context. *Strategic Change*, Vol. 16, Nos. 1–2, 69–78.

Zagurski, Tyler J. (2004). *Direct, Plan, or Influence? Joint C2 on the Future Battlefield*. Quantico, VA: U.S. Marine Corps, Marine Corps University, School of Advanced Warfighting.

Zhu, H., and R.Y. Wang (2010). Information Quality Framework for Verifiable Intelligence Products. In Y. Chan and T.M.M. Talley (Eds.), *Data Engineering: International Series in Operations Research and Management Science*. New York: Springer. pp. 1–19.

Zwikael, Ofer (2007). Al Qaeda's Operations: Project Management Analysis. *Studies in Conflict and Terrorism*, Vol. 30, No. 3, 267–280.

Index